W9-CSI-193

Statistics

Statistics

An Introduction using R

Michael J. Crawley
Imperial College London, UK

John Wiley & Sons, Ltd

1

Fundamentals

The hardest part of any statistical work is getting started – and one of the hardest things about getting started is choosing the right kind of statistical analysis. The choice depends on the nature of your data and on the particular question you are trying to answer. The truth is that there is no substitute for experience; the way to know what to do, is to have done it properly lots of times before.

The key is to understand what kind of **response** variable you have got, and to know the nature of your **explanatory** variables. The response variable is the thing you are working on; it is the variable whose variation you are attempting to understand. This is the variable that goes on the y axis of the graph (the ordinate). The explanatory variable goes on the x axis of the graph (the abscissa); you are interested in the extent to which variation in the response variable is associated with variation in the explanatory variable. A continuous measurement is a variable like height or weight that can take any real numbered value. A categorical variable is a **factor** with two or more **levels**: gender is a factor with two levels (male and female), and a rainbow might be a factor with seven levels (red, orange, yellow, green, blue, indigo, violet).

It is essential, therefore, that you know:

- which of your variables is the response variable;

- which are the explanatory variables;

- are the explanatory variables continuous or categorical, or a mixture of both;

- what kind of response variable have you got – is it a continuous measurement, a count, a proportion, a time-at-death or a category?

These simple keys will then lead you to the appropriate statistical method.

1. The explanatory variables

(a)	All explanatory variables continuous	**Regression**
(b)	All explanatory variables categorical	**Analysis of variance (Anova)**
(c)	Explanatory variables both continuous and categorical	**Analysis of covariance (Ancova)**

Statistics: An Introduction using R M. J. Crawley
© 2005 John Wiley & Sons, Ltd ISBNs: 0-470-02298-1 (PBK); 0-470-02297-3 (PPC)

2.   *The response variable*

(a)	Continuous	**Normal regression, Anova or Ancova**
(b)	Proportion	**Logistic regression**
(c)	Count	**Log linear models**
(d)	Binary	**Binary logistic analysis**
(e)	Time-at-death	**Survival analysis**

There are some key ideas that need to be understood from the outset. We cover these here before getting into any detail about different kinds of statistical model.

Everything Varies

If you measure the same thing twice you will get two different answers. If you measure the same thing on different occasions you will get different answers because the thing will have aged. If you measure different individuals, they will differ for both genetic and environmental reasons (nature and nurture). Heterogeneity is universal: spatial hetero-geneity means that places always differ and temporal heterogeneity means that times always differ.

Because everything varies, finding that things vary is simply not interesting. We need a way of discriminating between variation that is scientifically interesting, and variation that just reflects background heterogeneity. That is why we need statistics. It is what this whole book is about.

The key concept is the amount of variation that we would expect to occur by chance alone, when nothing scientifically interesting was going on. If we measure bigger differences than we would expect by chance, we say that the result is statistically significant. If we measure no more variation than we might reasonably expect to occur by chance alone, then we say that our result is not statistically significant. It is important to understand that this is not to say that the result is not important. Non-significant differences in human life span between two drug treatments may be massively important (especially if you are the patient involved). Non-significance is not the same as 'not different'. The lack of significance may simply be due to the fact that our replication is too low.

On the other hand, when nothing really **is** going on, then we want to know this. It makes life much simpler if we can be reasonably sure that there is no relationship between y and x. Some students think that 'the only good result is a significant result'. They feel that their study has somehow failed if it shows that 'A has no significant effect on B'. This is an understandable failing of human nature, but it is not good science. The point is that we want to know the truth, one way or the other. We should try not to care too much about the way things turn out. This is not an amoral stance, it just happens to be the way that science works best. Of course, it is hopelessly idealistic to pretend that this is the way that scientists really behave. Scientists often hope passionately that a particular experimental result will turn out to be statistically significant, so that they can have a paper published in Nature and get promoted, but that doesn't make it right.

Significance

What do we mean when we say that a result is significant? The normal dictionary definitions of significant are 'having or conveying a meaning' or 'expressive; suggesting or implying deeper or unstated meaning' but in statistics we mean something very specific indeed. We mean that 'a result was unlikely to have occurred by chance'. In particular, we mean 'unlikely to have occurred by chance if the null hypothesis was true'. So there are two elements to it: we need to be clear about what we mean by 'unlikely', and also what exactly we mean by the 'null hypothesis'. Statisticians have an agreed convention about what constitutes 'unlikely'. They say that an event is unlikely if it occurs less than 5% of the time. In general, the 'null hypothesis' says that 'nothing's happening' and the alternative says 'something **is** happening'.

Good and Bad Hypotheses

Karl Popper was the first to point out that a good hypothesis is one that is capable of **rejection**. He argued that **a good hypothesis is a falsifiable hypothesis**. Consider the following two assertions.

1. There are vultures in the local park.

2. There are no vultures in the local park.

Both involve the same essential idea, but one is refutable and the other is not. Ask yourself how you would refute option 1. You go out into the park and you look for vultures, but you don't see any. Of course, this doesn't mean that there aren't any. They could have seen you coming, and hidden behind you. No matter how long or how hard you look, you cannot refute the hypothesis. All you can say is 'I went out and I didn't see any vultures'. One of the most important scientific notions is that **absence of evidence is not evidence of absence**. Option 2 is fundamentally different. You reject hypothesis 2 the first time that you see a vulture in the park. Until the time that you **do** see your first vulture in the park, you work on the assumption that the hypothesis is true. But if you see a vulture, the hypothesis is clearly false, so you reject it.

Null Hypotheses

The null hypothesis says 'nothing's happening'. For instance, when we are comparing two sample means, the null hypothesis is that the means of the two samples are the same. Again, when working with a graph of y against x in a regression study, the null hypothesis is that the slope of the relationship is zero, i.e. y is not a function of x, or y is independent of x. The essential point is that the null hypothesis is falsifiable. We reject the null hypothesis when our data show that the null hypothesis is sufficiently unlikely.

p Values

A p value is an estimate of the probability that a particular result, or a result more extreme than the result observed, could have occurred by chance, if the null hypothesis were true. In short, the p value is a measure of the credibility of the null hypothesis. If something is very unlikely to have occurred by chance, we say that it is statistically significant, e.g. $p < 0.001$. For example, in comparing two sample means, where the null hypothesis is that the means are the same, a low p value means that the hypothesis is unlikely to be true and the difference is statistically significant. A large p value (e.g. $p = 0.23$) means that there is no compelling evidence on which to reject the null hypothesis. Of course, saying 'we do not reject the null hypothesis' and 'the null hypothesis is true' are two quite different things. For instance, we may have failed to reject a false null hypothesis because our sample size was too low, or because our measurement error was too large. Thus, p values are interesting, but they don't tell the whole story; effect sizes and sample sizes are equally important in drawing conclusions.

Interpretation

It should be clear by this point that we can make two kinds of mistakes in the interpretation of our statistical models:

- we can reject the null hypothesis when it is true, or
- we can accept the null hypothesis when it is false.

These are referred to as **Type I** and **Type II** errors respectively. Supposing we knew the true state of affairs (which, of course, we seldom do), then in tabular form:

	Actual situation	
Null hypothesis	*True*	*False*
Accept	Correct decision	Type II
Reject	Type I	Correct decision

Statistical Modelling

The object is to determine the values of the parameters in a specific model that lead to the best fit of the model to the data. The data are sacrosanct, and they tell us what actually happened under a given set of circumstances. It is a common mistake to say 'the data were fitted to the model' as if the data were something flexible, and we had a clear picture of the structure of the model. On the contrary, what we are looking for is the minimal adequate model to describe the data. The model is fitted to the data, not the other way around. The best model is the model that produces the least unexplained variation (the **minimal residual deviance**), subject to the constraint that all the parameters in the model should be statistically significant.

You have to specify the model. It embodies your mechanistic understanding of the factors involved, and of the way that they are related to the response variable. We want the model to be **minimal** because of the principle of parsimony, and **adequate** because there is no point in retaining an inadequate model that does not describe a significant fraction of the variation in the data. It is very important to understand that there is not just one model; this is one of the common implicit errors involved in traditional regression and Anova, where the same models are used, often uncritically, over and over again. In most circumstances, there will be a large number of different, more or less plausible models that might be fitted to any given set of data. Part of the job of data analysis is to determine which, if any, of the possible models are adequate and then, out of the set of adequate models, which is the minimal adequate model. In some cases there may be no single best model and a set of different models may all describe the data equally well (or equally poorly if the variability is great).

Maximum Likelihood

What exactly do we mean when we say that the parameter values should afford the 'best fit of the model to the data'? The convention we adopt is that our techniques should lead to **unbiased, variance minimizing estimators**. We define 'best' in terms of **maximum likelihood**. This notion is likely to be unfamiliar, so it is worth investing some time to get a feel for it. This is how it works.

- Given the data,

- and given our choice of model,

- what values of the parameters of that model make the observed data most likely?

Here are the data: y is the response variable and x is the explanatory variable. Because both x and y are continuous variables, the appropriate model is regression.

```
x<-c(1,3,4,6,8,9,12)
y<-c(5,8,6,10,9,13,12)
plot(x,y)
```

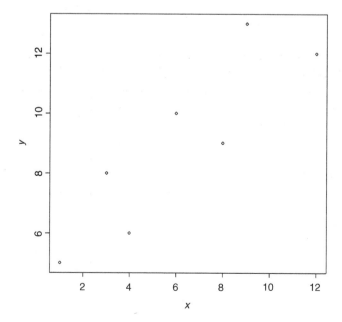

Now we need to select a regression model to describe these data from the vast range of possible models available. Let's choose the simplest model, the straight line

$$y = a + bx.$$

This is a two-parameter model; the first parameter, a, is the intercept (the value of y when x is 0) and the second, b, is the slope (the change in y associated with unit change in x). The response variable y, is a linear function of the explanatory variable x. Now suppose that we knew that the slope was 0.68, then the maximum likelihood question can be applied to the intercept a.

If the intercept were 0 (left-hand graph, below), would the data be likely? The answer of course, is no. If the intercept were 8 (right-hand graph) would the data be likely? Again, the answer is obviously no. The maximum likelihood estimate of the intercept is shown in the central graph (its value turns out to be 4.827).

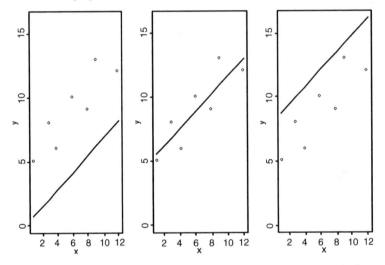

We could have a similar debate about the slope. Suppose we knew that the intercept was 4.827, then would the data be likely if the graph had a slope of 1.5 (left graph, below)?

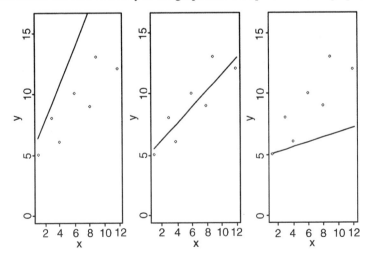

The answer, of course, is no. What about a slope of 0.2 (right graph)? Again, the data are not at all likely if the graph has such a gentle slope. The maximum likelihood of the data given the model is obtained with a slope of 0.679 (centre graph). This is not how the procedure is actually carried out, but it makes the point that we judge the model on the basis of how likely the data would be if the model were correct. In practice of course, both parameters are estimated simultaneously.

Experimental Design

There are only two key concepts:

- replication, and

- randomization.

You replicate to increase reliability. You randomize to reduce bias. If you replicate thoroughly and randomize properly, you will not go far wrong.

There are a number of other issues whose mastery will increase the likelihood that you analyse your data the right way rather than the wrong way:

- the principle of parsimony,

- the power of a statistical test,

- controls,

- spotting pseudoreplication and knowing what to do about it,

- the difference between experimental and observational data (non-orthogonality).

It does not matter very much if you cannot do your own advanced statistical analysis. If your experiment is properly designed, you will often be able to find somebody to help you with the statistics. However, if your experiment is not properly designed, or not thoroughly randomized, or lacking adequate controls, then no matter how good you are at statistics, some (or possibly even all) of your experimental effort will have been wasted. No amount of high-powered statistical analysis can turn a bad experiment into a good one. R is good, but not that good.

The Principle of Parsimony (Occam's Razor)

One of the most important themes running through this book concerns model simplification. The principle of parsimony is attributed to the 14th century English Nominalist philosopher William of Occam who insisted that, given a set of equally good explanations for a given phenomenon, then **the correct explanation is the simplest explanation**. It is called Occam's razor because he 'shaved' his explanations down to the bare minimum. In statistical modelling, the principle of parsimony means that:

- models should have as few parameters as possible,

- linear models should be preferred to non-linear models,

- experiments relying on few assumptions should be preferred to those relying on many,

- models should be pared down until they are *minimal adequate*,

- simple explanations should be preferred to complex explanations.

The process of model simplification is an integral part of hypothesis testing in R. In general, a variable is retained in the model only if it causes a significant increase in deviance when it is removed from the current model. Seek simplicity, then distrust it.

In our zeal for model simplification, we must be careful not to throw the baby out with the bathwater. Einstein made a characteristically subtle modification to Occam's razor. He said: 'A model should be as simple as possible. But no simpler'.

Observation, Theory and Experiment

There is no doubt that the best way to solve scientific problems is through a thoughtful blend of observation, theory and experiment. In most real situations, however, there are constraints on what can be done, and on the way things can be done, which mean that one or more of the trilogy has to be sacrificed. There are lots of cases, for example, where it is ethically or logistically impossible to carry out manipulative experiments. In these cases it is doubly important to ensure that the statistical analysis leads to conclusions that are as critical and as unambiguous as possible.

Controls

No controls, no conclusions.

Replication: It's the *n*'s that Justify the Means

The requirement for replication arises because if we do the same thing to different individuals we are likely to get different responses. The causes of this heterogeneity in response are many and varied (genotype, age, gender, condition, history, substrate, microclimate, and so on). The object of replication is to increase the reliability of parameter estimates, and to allow us to quantify the variability that is found within the same treatment. To qualify as replicates, the repeated measurements:

- must be independent,

- must not form part of a time series (data collected from the same place on successive occasions are not independent),

- must not be grouped together in one place (aggregating the replicates means that they are not spatially independent),

- must be of an appropriate spatial scale.

Ideally, one replicate from each treatment ought to be grouped together into a block, and each treatment repeated in many different blocks. Repeated measures (e.g. from the same individual or the same spatial location) are not replicates (this is probably the commonest cause of pseudoreplication in statistical work).

How Many Replicates?

The usual answer is 'as many as you can afford'. An alternative answer is 30. A very useful rule of thumb is this: a sample of 30 or more is a big sample, but a sample of less than 30 is a small one. The rule doesn't always work, of course: 30 would be derisively small as a sample in an opinion poll, for instance. In other circumstances, it might be impossibly expensive to repeat an experiment as many as 30 times. Nevertheless, it is a rule of great practical utility, if only for giving you pause as you design your experiment with 300 replicates that perhaps this might really be a bit over the top – or when you think you could get away with just five replicates this time.

There are ways of working out the replication necessary for testing a given hypothesis (these are explained below). Sometimes we know little or nothing about the variance or the response variable when we are planning an experiment. Experience is important. So are pilot studies. These should give an indication of the variance between initial units before the experimental treatments are applied, and also of the approximate magnitude of the responses to experimental treatment that are likely to occur. Sometimes it may be necessary to reduce the scope and complexity of the experiment, and to concentrate the inevitably limited resources of manpower and money on obtaining an unambiguous answer to a simpler question. It is immensely irritating to spend three years on a grand experiment, only to find at the end of it that the response is only significant at $p = 0.08$. A reduction in the number of treatments might well have allowed an increase in replication to the point where the same result would have been unambiguously significant.

Power

The power of a test is the probability of rejecting the null hypothesis when it is false. It has to do with Type II errors: β is the probability of accepting the null hypothesis when it is false. In an ideal world, we would obviously make β as small as possible, but there is a snag. The smaller we make the probability of committing a Type II error, the greater we make the probability of committing a Type I error, and rejecting the null hypothesis when, in fact, it is correct. A compromise is called for. Most statisticians work with $\alpha = 0.05$ and $\beta = 0.2$. Now the power of a test is defined as $1 - \beta = 0.8$ under the standard assumptions. This is used to calculate the sample sizes necessary to detect a specified difference when the error variance is known (or can be guessed at). Suppose that for a single sample the size of the difference you want to detect is ∂ and the variance in the response is s^2 (e.g. known from a pilot study or extracted from the literature), then you will need n replicates to reject the null hypothesis with power $= 80\%$:

$$n \approx \frac{8 \times s^2}{\partial^2}.$$

This is a reasonable rule of thumb, but you should err on the side of caution by having larger, not smaller samples than these. Suppose that the mean is close to 20, and the variance is 10, but we want to detect a 10% change (i.e. $\partial = \pm 2$) with probability 0.8, then $n = 8 \times 10/2^2 = 20$.

Here is the built-in function **power.t.test** in action for the case just considered. We need to specify that the type is "one sample", the power we want to obtain is 0.8, the difference to be detected (called delta) is 2.0, and the standard deviation (sd) is $\sqrt{10}$

power.t.test(type = "one.sample",power = 0.8,sd = sqrt(10),delta = 2)

```
      One-sample t test power calculation

            n = 21.62146
        delta = 2
           sd = 3.162278
    sig.level = 0.05
        power = 0.8
  alternative = two.sided
```

Other power functions available in R include **power.anova.test** and **power.prop.test**

Randomization

Randomization is something that everybody says they do, but hardly anybody does properly. Take a simple example. How do I select one tree from a forest of trees, on which to measure photosynthetic rates? I want to select the tree at random in order to avoid bias. For instance, I might be tempted to work on a tree that had accessible foliage near to the ground, or a tree that was close to the lab, or a tree that looked healthy, or a tree that had nice insect-free leaves, and so on. I leave it to you to list the biases that would be involved in estimating photosynthesis on any of those trees. One common way of selecting a 'random' tree is to take a map of the forest and select a random pair of coordinates (say 157m east of the reference point, and 68m north). Then pace out these coordinates and, having arrived at that particular spot in the forest, select the nearest tree to those coordinates. But is this really a randomly selected tree?

If it was randomly selected, then it would have exactly the same chance of being selected as every other tree in the forest. Let us think about this. Look at the figure below which shows a plan of the distribution of trees on the ground. Even if they were originally planted out in regular rows, accidents, tree-falls, and heterogeneity in the substrate would soon lead to an aggregated spatial distribution of trees. Now ask yourself how many different random points would lead to the selection of a given tree. Start with tree (a). This will be selected by any points falling in the large shaded area.

Now consider tree (b). It will only be selected if the random point falls within the tiny area surrounding that tree. Tree (a) has a much greater chance of being selected than tree (b), and so the nearest tree to a random point is not a randomly selected tree. In a spatially heterogeneous woodland, isolated trees and trees on the edges of clumps will always have a higher probability of being picked than trees in the centre of clumps.

The answer is that to select a tree at random, every single tree in the forest must be numbered (all 24 683 of them, or whatever), and then a random number between 1 and 24 683 must be drawn out of a hat. There is no alternative. Anything less than that is not randomization.

Now ask yourself how often this is done in practice, and you will see what I mean when I say that randomization is a classic example of 'do as I say, and not do as I do'. As

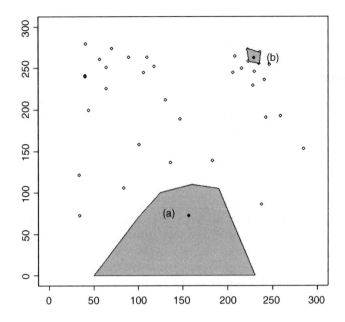

an example of how important proper randomization can be, consider the following experiment that was designed to test the toxicity of five contact insecticides by exposing batches of flour beetles to the chemical on filter papers in Petri dishes. The animals walk about and pick up the poison on their feet. The *Tribolium* culture jar was inverted, flour and all, into a large tray, and beetles were collected as they emerged from the flour. The animals were allocated to the five chemicals in sequence; four replicate Petri dishes were treated with the first chemical, and ten beetles were placed in each Petri dish. Do you see the source of bias in this procedure?

It is entirely plausible that flour beetles differ in their activity levels (gender differences, differences in body weight, age, etc.). The most active beetles might emerge first from the pile of flour. These beetles all end up in the treatment with the first insecticide. By the time we come to finding beetles for the last replicate of the fifth pesticide, we may be grubbing round in the centre of the pile, looking for the last remaining *Tribolium*. This matters, because the amount of pesticide picked up by the beetles will depend upon their activity levels. The more active the beetles, the more chemical they pick up, and the more likely they are to die. Thus, the failure to randomize will bias the result in favour of the first insecticide because this treatment received the most active beetles.

What we should have done is this. Fill $5 \times 4 = 20$ Petri dishes with ten beetles each, adding one beetle to each Petri dish in turn. Then allocate a treatment (one of the five pesticides) to each Petri dish at random, and place the beetles on top of the pre-treated filter paper. We allocate Petri dishes to treatments most simply by writing a treatment number of a slip of paper, and placing all 20 pieces of paper in a bag. Then draw one piece of paper from the bag. This gives the treatment number to be allocated to the Petri dish in question. All of this may sound absurdly long-winded but, believe me, it is vital.

The recent trend towards 'haphazard' sampling is a cop-out. What it means is that 'I admit that I didn't randomize, but you have to take my word for it that this did not introduce any important bias'. You can draw your own conclusions.

Strong Inference

One of the most powerful means available to demonstrate the accuracy of an idea is an experimental confirmation of a prediction made by a carefully formulated hypothesis. There are two essential steps to the protocol of strong inference (Platt 1964):

* formulate a clear hypothesis, and

* devise an acceptable test.

Neither one is much good without the other. For example, the hypothesis should not lead to predictions that are likely to occur by other extrinsic means. Similarly, the test should demonstrate unequivocally whether the hypothesis is true or false.

A great many scientific experiments appear to be carried out with no particular hypothesis in mind at all, but simply to see what happens. While this approach may be commendable in the early stages of a study, such experiments tend to be weak as an end in themselves, because there will be such a large number of equally plausible explanations for the results. Without contemplation there will be no testable predictions; without testable predictions there will be no experimental ingenuity; without experimental ingenuity there is likely to be inadequate control; in short, equivocal interpretation. The results could be due to myriad plausible causes. Nature has no stake in being understood by scientists. We need to work at it. Without replication, randomization and good controls we shall make little progress.

Weak Inference

The phrase weak inference is used (often disparagingly) to describe the interpretation of observational studies and the analysis of so-called 'natural experiments'. It is silly to be disparaging about these data, because they are often the only data that we have. The aim of good statistical analysis is to obtain the maximum information from a given set of data, bearing the limitations of the data firmly in mind.

Natural experiments arise when an event (often assumed to be an unusual event, but frequently without much justification of what constitutes unusualness) occurs that is like an experimental treatment (a hurricane blows down half of a forest block; a landslide creates a bare substrate; a stock market crash produces lots of suddenly poor people, etc). Hairston (1989) said: 'The requirement of adequate knowledge of initial conditions has important implications for the validity of many natural experiments. Inasmuch as the "experiments" are recognized only when they are completed, or in progress at the earliest, it is impossible to be certain of the conditions that existed before such an "experiment" began. It then becomes necessary to make assumptions about these conditions, and any conclusions reached on the basis of natural experiments are thereby weakened to the point of being hypotheses, and they should be stated as such' (Hairston 1989).

How Long to Go On?

Ideally, the duration of an experiment should be determined in advance, lest one falls prey to one of the twin temptations:

- to stop the experiment as soon as a pleasing result is obtained;

- to keep going with the experiment until the 'right' result is achieved (the 'Gregor Mendel effect').

In practice, most experiments probably run for too short a period, because of the idiosyncrasies of scientific funding. This short-term work is particularly dangerous in medicine and the environmental sciences, because the kind of short-term dynamics exhibited after pulse experiments may be entirely different from the long-term dynamics of the same system. Only by long-term experiments of both the pulse and the press kind, will the full range of dynamics be understood. The other great advantage of long-term experiments is that a wide range of patterns (e.g. 'kinds of years') is experienced.

Pseudoreplication

Pseudoreplication occurs when you analyse the data as if you had more degrees of freedom than you really have. There are two kinds of pseudoreplication:

- temporal pseudoreplication, involving repeated measurements from the same individual, and

- spatial pseudoreplication, involving several measurements taken from the same vicinity.

Pseudoreplication is a problem because one of the most important assumptions of standard statistical analysis is **independence of errors**. Repeated measures through time on the same individual will have non-independent errors because peculiarities of the individual will be reflected in all of the measurement made on it (the repeated measures will be temporally correlated with one another). Samples taken from the same vicinity will have non-independent errors because peculiarities of the location will be common to all the samples (e.g. yields will all be high in a good patch and all be low in a bad patch).

Pseudoreplication is generally quite easy to spot. The question to ask is how many degrees of freedom for error does the experiment really have? If a field experiment appears to have lots of degrees of freedom, it is probably pseudoreplicated. Take an example from pest control of insects on plants. There are 20 plots, ten sprayed and ten unsprayed. Within each plot there are 50 plants. Each plant is measured five times during the growing season. Now this experiment generates $20 \times 50 \times 5 = 5000$ numbers. There are two spraying treatments, so there must be 1 degree of freedom for spraying and 4998 degrees of freedom for error. Or must there? Count up the replicates in this experiment. Repeated measurements on the same plants (the five sampling occasions) are certainly not replicates. The 50 individual plants within each quadrat are not replicates either. The reason for this is that conditions within each quadrat are quite likely to be unique, and so all 50 plants will experience more or less the same unique set of conditions, irrespective of the spraying treatment they receive. In fact, there are ten replicates in this experiment. There are ten sprayed plots and ten unsprayed plots, and each plot will yield only one independent datum to the response variable (the proportion of leaf area consumed by insects, for example). Thus, there are nine degrees of freedom within each treatment, and $2 \times 9 = 18$ degrees of freedom for error in the experiment as a whole. It is not difficult to find examples of pseudoreplication on this scale in the literature (Hurlbert 1984). The problem is that it leads to the reporting of masses of spuriously significant results (with 4998 degrees of freedom for error, it is almost impossible not to have significant differences). The first skill to be acquired by the budding experimenter is the ability to plan an experiment that is properly replicated.

There are various things that you can do when your data are pseudoreplicated:

- average away the pseudoreplication and carry out your statistical analysis on the means,

- carry out separate analyses for each time period,

- use proper time series analysis or mixed effects models.

Initial Conditions

Many otherwise excellent scientific experiments are spoiled by a lack of information about initial conditions. How can we know if something has changed if we don't know what it was like to begin with? It is often implicitly assumed that all the experimental units were alike at the beginning of the experiment, but this needs to be demonstrated rather than taken on faith. One of the most important uses of data on initial conditions is as a check on the efficiency of randomization. For example, you should be able to run your statistical analysis to demonstrate that the individual organisms were not significantly different in mean size at the beginning of a growth experiment. Without measurements of initial size, it is always possible to attribute the end result to differences in initial conditions. Another reason for measuring initial conditions is that the information can often be used to improve the resolution of the final analysis through analysis of covariance (see Chapter 10).

Orthogonal Designs and Non-orthogonal Observational Data

The data in this book fall into two distinct categories. In the case of planned experiments, all of the treatment combinations are equally represented and, barring accidents, there are no missing values. Such experiments are said to be *orthogonal*. In the case of observational studies, however, we have no control over the number of individuals for which we have data, or over the combinations of circumstances that are observed. Many of the explanatory variables are likely to be correlated with one another, as well as with the response variable. Missing treatment combinations are commonplace, and the data are said to be non-orthogonal. This makes an important difference to our statistical modelling because, in orthogonal designs, the deviance that is attributed to a given factor is constant, and does not depend upon the order in which that factor is removed from the model. In contrast, with non-orthogonal data, we find that the deviance attributable to a given factor does depend upon the order in which the factor is removed from the model. We must be careful, therefore, to judge the significance of factors in non-orthogonal studies, when they are removed from the maximal model (i.e. from the model including all the other factors and interactions with which they might be confounded). Remember, for non-orthogonal data, order matters.

2

Dataframes

Learning how to handle your data, how to enter it into the computer, and how to read the data into R are amongst the most important topics you will need to master. R handles data in objects known as dataframes. A dataframe is an object with rows and columns (a bit like a two-dimensional matrix). The rows contain different observations from your study, or measurements from your experiment. The columns contain the values of different variables. The values in the body of the dataframe can be numbers (as they would be in as matrix), but they could also be text (e.g. the names of factor levels for categorical variables, like 'male' or 'female' in a variable called 'gender'), they could be calendar dates (like 23/5/04), or they could be logical variables (like 'true' or 'false'). Here is a spreadsheet in the form of a dataframe with seven variables, the left-most of which comprises the row names, and other variables are numeric (area, slope, soil pH and worm density), categorical (field name and vegetation) or logical (damp is either true = T or false = F).

Field name	Area	Slope	Vegetation	Soil pH	Damp	Worm density
Nash's Field	3.6	11	Grassland	4.1	F	4
Silwood Bottom	5.1	2	Arable	5.2	F	7
Nursery Field	2.8	3	Grassland	4.3	F	2
Rush Meadow	2.4	5	Meadow	4.9	T	5
Gunness' Thicket	3.8	0	Scrub	4.2	F	6
Oak Mead	3.1	2	Grassland	3.9	F	2
Church Field	3.5	3	Grassland	4.2	F	3
Ashurst	2.1	0	Arable	4.8	F	4
The Orchard	1.9	0	Orchard	5.7	F	9
Rookery Slope	1.5	4	Grassland	5	T	7
Garden Wood	2.9	10	Scrub	5.2	F	8
North Gravel	3.3	1	Grassland	4.1	F	1
South Gravel	3.7	2	Grassland	4	F	2
Observatory Ridge	1.8	6	Grassland	3.8	F	0
Pond Field	4.1	0	Meadow	5	T	6
Water Meadow	3.9	0	Meadow	4.9	T	8

Statistics: An Introduction using R M. J. Crawley
© 2005 John Wiley & Sons, Ltd ISBNs: 0-470-02298-1 (PBK); 0-470-02297-3 (PPC)

Field name	Area	Slope	Vegetation	Soil pH	Damp	Worm density
Cheapside	2.2	8	Scrub	4.7	T	4
Pound Hill	4.4	2	Arable	4.5	F	5
Gravel Pit	2.9	1	Grassland	3.5	F	1
Farm Wood	0.8	10	Scrub	5.1	T	3

Perhaps the most important thing about analysing your own data properly is getting your dataframe absolutely right. The expectation is that you will have used a spreadsheet like Excel to enter and edit the data, and that you will have used plots to check for errors. The thing that takes some practice is learning exactly how to put your numbers into the spreadsheet. There are countless ways of doing it wrong, but only one way of doing it right – and this way is not the way that most people find intuitively to be the most obvious.

The key thing is this: **all the values of the same variable must go in the same column**. It does not sound like much, but this is what people tend to get wrong. If you had an experiment with three treatments (control, pre-heated and pre-chilled), and four measurements per treatment, it might seem like a good idea to create the spreadsheet like this:

Control	Pre-heated	Pre-chilled
6.1	6.3	7.1
5.9	6.2	8.2
5.8	5.8	7.3
5.4	6.3	6.9

However, this is not a dataframe, because values of the response variable appear in three different columns, rather than all in the same column. The correct way to enter these data is to have two columns: one for the response variable and one for the levels of the experimental factor (control, pre-heated and pre-chilled). Here are the same data, entered correctly as a dataframe

Response	Treatment
6.1	Control
5.9	Control
5.8	Control
5.4	Control
6.3	Pre-heated
6.2	Pre-heated
5.8	Pre-heated
6.3	Pre-heated
7.1	Pre-chilled
8.2	Pre-chilled
7.3	Pre-chilled
6.9	Pre-chilled

A good way to practice this layout is to use the Excel function called Pivot Table (found under the Data tab on the main menu bar) on your own data; it requires your spreadsheet to be in the form of a dataframe, with each of the explanatory variables in its own column.

Once you have made your dataframe in Excel and corrected all the inevitable data-entry and spelling errors, then you need to save the dataframe in a file format that can be read by R. Much the simplest way is to save all your dataframes from Excel as tab-delimited text files: File/Save As/... then from the 'Save as type' options choose 'Text (Tab delimited)'. There is no need to add a suffix, because Excel will automatically add '.txt' to your file name. This file can then be read into R directly as a dataframe, using the read.table function.

It is important to note that read.table would fail if there were any spaces in any of the variable names in row 1 of the dataframe (the header row) like Field name, Soil pH or Worm density, or between any of the words within the same factor level (as in many of the Field names). We should replace all these spaces by dots '.' before saving the dataframe in Excel (use Edit/Replace with " " replaced by "."). Now the dataframe can be read into R. There are three things to remember:

- the whole path and file name needs to be enclosed in double quotes: "c:\\abc.txt",

- header = T says that the first row contains the variable names,

- always use double backslash \\ rather than \ in the file path definition.

Think of a name for the data frame (say 'worms' in this case). Now use the **gets arrow** < − which is a composite symbol made up of the two characters < (less than) and − (minus) like this

worms < -read.table("c:\\temp\\worms.txt",header = T,row.names = 1)

Once the file has been imported to R we want to do two things:

- use attach to make the variables accessible by name within the R session, and

- use names to get a list of the variable names.

Typically, the two commands are issued in sequence, whenever a new dataframe is imported from file:

```
attach(worms)
names(worms)
[ 1]   Field.Name"   "Area"   "Slope"        "Vegetation"
[ 5]   "Soil.pH"   "Damp"   "Worm.density"
```

To see the contents of the dataframe, just type its name

worms

	Area	Slope	Vegetation	Soil.pH	Damp	Worm.density
Nash's.Field	3.6	11	Grassland	4.1	FALSE	4
Silwood.Bottom	5.1	2	Arable	5.2	FALSE	7
Nursery.Field	2.8	3	Grassland	4.3	FALSE	2
Rush.Meadow	2.4	5	Meadow	4.9	TRUE	5
Gunness'.Thicket	3.8	0	Scrub	4.2	FALSE	6
Oak.Mead	3.1	2	Grassland	3.9	FALSE	2
Church.Field	3.5	3	Grassland	4.2	FALSE	3
Ashurst	2.1	0	Arable	4.8	FALSE	4
The.Orchard	1.9	0	Orchard	5.7	FALSE	9
Rookery.Slope	1.5	4	Grassland	5.0	TRUE	7
Garden.Wood	2.9	10	Scrub	5.2	FALSE	8
North.Gravel	3.3	1	Grassland	4.1	FALSE	1
South.Gravel	3.7	2	Grassland	4.0	FALSE	2
Observatory.Ridge	1.8	6	Grassland	3.8	FALSE	0
Pond.Field	4.1	0	Meadow	5.0	TRUE	6
Water.Meadow	3.9	0	Meadow	4.9	TRUE	8
Cheapside	2.2	8	Scrub	4.7	TRUE	4
Pound.Hill	4.4	2	Arable	4.5	FALSE	5
Gravel.Pit	2.9	1	Grassland	3.5	FALSE	1
Farm.Wood	0.8	10	Scrub	5.1	TRUE	3

If, as here, the rows have unique names as part of the dataframe, we suppress R's natural inclination to produce its own row numbers, by telling R the number of the column containing the row names (1 in this case) as part of the read.table function (see above). Notice, also, that R has expanded our abbreviated T and F into TRUE and FALSE. The object called worms now has all the attributes of a dataframe. For example, you can summarize it, using **summary**:

summary(worms)

Area	Slope	Vegetation	Soil.pH	Damp
Min. : 0.800	Min. : 0.00	Arable : 3	Min. : 3.500	Mode :logical
1st Qu.: 2.175	1st Qu.: 0.75	Grassland : 9	1st Qu.: 4.100	FALSE : 14
Median : 3.000	Median : 2.00	Meadow : 3	Median : 4.600	TRUE : 6
Mean : 2.990	Mean : 3.50	Orchard : 1	Mean : 4.555	
3rd Qu.: 3.725	3rd Qu.: 5.25	Scrub : 4	3rd Qu.: 5.000	
Max. : 5.100	Max. : 11.00		Max. : 5.700	

```
Worm.density
Min.   : 0.00
1st Qu.: 2.00
Median : 4.00
Mean   : 4.35
3rd Qu.: 6.25
Max.   : 9.00
```

Values of continuous variables are summarized under six headings: one parametric (the arithmetic mean) and five non-parametric (maximum, minimum, median, 25 percentile – the first quartile – and 75 percentile – the third quartile). Levels of categorical variables are counted. Note that the field names are not summarized, because they have been declared to be the row names, and hence all the names have to be unique.

Selecting Parts of a Dataframe: Subscripts

We often want to extract part of a dataframe. This is a very general procedure in R, accomplished using what are called subscripts. You can think of subscripts as addresses within a vector, a matrix or a dataframe. Subscripts in R appear within square brackets, thus y[7] is the seventh element of the vector called y and z[2,6] is the second row of the sixth column of a two-dimensional matrix called z. This is in contrast to arguments to functions in R, which appear in round brackets (4,7).

We might want to select all the rows of a dataframe for certain specified columns. Or we might want to select all the columns for certain specified rows of the dataframe. The convention in R is that when we do not specify any subscript, then all the rows, or all the columns is assumed. This syntax is difficult to understand on first acquaintance, but [,'blank then comma' means 'all the rows' and,] 'comma then blank' means all the columns. For instance, to select the first column of the dataframe, use subscript [,1]. Thus, to select all the rows of the first three columns of worms, we write:

worms[,1:3]

	Area	Slope	Vegetation
Nash's.Field	3.6	11	Grassland
Silwood.Bottom	5.1	2	Arable
Nursery.Field	2.8	3	Grassland
Rush.Meadow	2.4	5	Meadow
Gunness'.Thicket	3.8	0	Scrub
Oak.Mead	3.1	2	Grassland
Church.Field	3.5	3	Grassland
Ashurst	2.1	0	Arable
The.Orchard	1.9	0	Orchard
Rookery.Slope	1.5	4	Grassland
Garden.Wood	2.9	10	Scrub
North.Gravel	3.3	1	Grassland
South.Gravel	3.7	2	Grassland
Observatory.Ridge	1.8	6	Grassland
Pond.Field	4.1	0	Meadow
Water.Meadow	3.9	0	Meadow
Cheapside	2.2	8	Scrub
Pound.Hill	4.4	2	Arable
Gravel.Pit	2.9	1	Grassland
Farm.Wood	0.8	10	Scrub

To select just the middle 11 rows for all the columns of the dataframe, use subscript [5:15,] like this:

worms[5:15,]

	Area	Slope	Vegetation	Soil.pH	Damp	Worm.density
Gunness'.Thicket	3.8	0	Scrub	4.2	FALSE	6
Oak.Mead	3.1	2	Grassland	3.9	FALSE	2
Church.Field	3.5	3	Grassland	4.2	FALSE	3
Ashurst	2.1	0	Arable	4.8	FALSE	4
The.Orchard	1.9	0	Orchard	5.7	FALSE	9
Rookery.Slope	1.5	4	Grassland	5.0	TRUE	7
Garden.Wood	2.9	10	Scrub	5.2	FALSE	8
North.Gravel	3.3	1	Grassland	4.1	FALSE	1
South.Gravel	3.7	2	Grassland	4.0	FALSE	2
Observatory.Ridge	1.8	6	Grassland	3.8	FALSE	0
Pond.Field	4.1	0	Meadow	5.0	TRUE	6

It is often useful to select certain rows, based on **logical tests** on the values of one or more variables. Here is the code to select only those rows which have Area > 3 and Slope < 3 using 'comma then blank' like this:

worms[Area > 3 & Slope < 3,]

	Area	Slope	Vegetation	Soil.pH	Damp	Worm.density
Silwood.Bottom	5.1	2	Arable	5.2	FALSE	7
Gunness'.Thicket	3.8	0	Scrub	4.2	FALSE	6
Oak.Mead	3.1	2	Grassland	3.9	FALSE	2
North.Gravel	3.3	1	Grassland	4.1	FALSE	1
South.Gravel	3.7	2	Grassland	4.0	FALSE	2
Pond.Field	4.1	0	Meadow	5.0	TRUE	6
Water.Meadow	3.9	0	Meadow	4.9	TRUE	8
Pound.Hill	4.4	2	Arable	4.5	FALSE	5

Sorting

You can sort the rows or the columns of the dataframe in any way you choose, but you need to state which columns you want to be sorted (typically you will want all of them sorted, i.e. columns 1:6 in the case of **worms**). Suppose we want the rows of the whole dataframe sorted by Area (this is the variable in column number one [,1]):

worms[order(worms[,1]),1:6]

	Area	Slope	Vegetation	Soil.pH	Damp	Worm.density
Farm.Wood	0.8	10	Scrub	5.1	TRUE	3
Rookery.Slope	1.5	4	Grassland	5.0	TRUE	7
Observatory.Ridge	1.8	6	Grassland	3.8	FALSE	0

The.Orchard	1.9	0	Orchard	5.7	FALSE	9
Ashurst	2.1	0	Arable	4.8	FALSE	4
Cheapside	2.2	8	Scrub	4.7	TRUE	4
Rush.Meadow	2.4	5	Meadow	4.9	TRUE	5
Nursery.Field	2.8	3	Grassland	4.3	FALSE	2
Garden.Wood	2.9	10	Scrub	5.2	FALSE	8
Gravel.Pit	2.9	1	Grassland	3.5	FALSE	1
Oak.Mead	3.1	2	Grassland	3.9	FALSE	2
North.Gravel	3.3	1	Grassland	4.1	FALSE	1
Church.Field	3.5	3	Grassland	4.2	FALSE	3
Nash's.Field	3.6	11	Grassland	4.1	FALSE	4
South.Gravel	3.7	2	Grassland	4.0	FALSE	2
Gunness'.Thicket	3.8	0	Scrub	4.2	FALSE	6
Water.Meadow	3.9	0	Meadow	4.9	TRUE	8
Pond.Field	4.1	0	Meadow	5.0	TRUE	6
Pound.Hill	4.4	2	Arable	4.5	FALSE	5
Silwood.Bottom	5.1	2	Arable	5.2	FALSE	7

Alternatively, the dataframe can be sorted in descending order by Soil pH, with only Soil pH and Worm density as output:

worms[rev(order(worms[,4])),c(4,6)]

	Soil.pH	Worm.density
The.Orchard	5.7	9
Garden.Wood	5.2	8
Silwood.Bottom	5.2	7
Farm.Wood	5.1	3
Pond.Field	5.0	6
Rookery.Slope	5.0	7
Water.Meadow	4.9	8
Rush.Meadow	4.9	5
Ashurst	4.8	4
Cheapside	4.7	4
Pound.Hill	4.5	5
Nursery.Field	4.3	2
Church.Field	4.2	3
Gunness'.Thicket	4.2	6
North.Gravel	4.1	1
Nash's.Field	4.1	4
South.Gravel	4.0	2
Oak.Mead	3.9	2
Observatory.Ridge	3.8	0
Gravel.Pit	3.5	1

Saving Your Work

At any stage, you can highlight material on the screen, then use copy (Ctrl C) and paste (Ctrl V) to save it to a Word document. Note that to keep tabular material properly aligned in the Word document you will need to use a font like Courier New that has absolute (rather than proportional) spacing. Graphs that you want to keep should be saved as you go along (using File: Save As when the graphics window is highlighted, to chose an appropriate format), or copied and pasted into a Word document.

You can review the command lines entered during a session with

history(Inf)

and you can copy from this and paste into the command line to save re-typing. You can save the history of command lines to a text file like this

savehistory("c:\\temp\\today.txt")

and read it back into R with loadhistory("c:\\temp\\today.txt"). The session as a whole can be saved as a binary file with

save(list = ls(), file = "c:\\temp\\all.Rdata")

and retrieved using load("c:\\temp\\all.Rdata")

Tidying Up

At the end of a session in R, it is good practice to remove (rm) any variables names you have created (using, say, x <- 5.6) and to detach any dataframes you have attached earlier in the session. That way, variables with the same names but different properties will not get in each other's way in subsequent work:

rm(x,y,z)
detach(worms)

This command does not make the dataframe called worms disappear; it just means that the variables within worms, like Slope and Area, are no longer accessible directly by name. To get rid of everything, including all the dataframes, type

rm(list = ls())

but be absolutely sure that you want to be as Draconian as this before you execute the command.

3

Central Tendency

Despite the fact that everything varies, measurements often cluster around certain intermediate values; this attribute is called central tendency. Even if the data themselves do not show much tendency to cluster round some central value, then parameters derived from repeated experiments (e.g. replicated sample means) almost inevitably do (this is called the central limit theorem; see p. 55). We need some data to work with:

```
yvals < -read.table("c:\\temp\\yvalues.txt",header = T)
attach(yvals)
```

So how should we quantify central tendency? Perhaps the most obvious way is just by looking at the data, without doing any calculations at all. The data values that occur most frequently are called the **mode**, and we discover the value of the mode simply by drawing a histogram of the data like this:

```
hist(y)
```

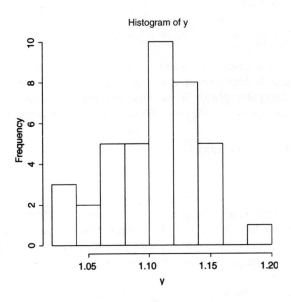

Statistics: An Introduction using R M. J. Crawley
© 2005 John Wiley & Sons, Ltd   ISBNs: 0-470-02298-1 (PBK); 0-470-02297-3 (PPC)

So we would say that the modal class of *y* was between 1.10 and 1.12 (we'll see how to control the location of the break points in a histogram later).

The most straightforward quantitative measure of central tendency is the **arithmetic mean** of the data. This is the sum of all the data values $\sum y$ divided by the number of data values, *n*. The capital Greek sigma $\sum$ just means 'add up all the values' of what follows; in this case, a set of *y* values. So if we call the arithmetic mean '*y* bar', $\bar{y}$, we can write

$$\bar{y} = \frac{\sum y}{n}.$$

The formula shows how we would write a general function to calculate arithmetic means for any vector of *y* values. First, we need to add them up. We could do it like this:

y[1] + y[2] + y[3] + ... y[n]

but that is very long-winded and it supposes that we know the value of *n* in advance. Fortunately, there is a built-in function called sum that works for any length of vector, so

total <- sum(y)

gives us the value for the numerator. Now what about the number of data values? This is likely to vary from application to application. We could print out the *y* values and count them, but that is very tedious and error-prone. There is a very important, general function in R to work this out for us. The function is called length(y) and it returns the number of numbers in the vector called *y*:

n <- length(y)

So our function for calculating the arithmetic mean would be ybar <- total/n. There is no need to calculate the intermediate values, *total* and *n*, so it would be more efficient to write ybar <- sum(y)/length(y). To put this logic into a general function we need to pick a name for the function, let's say 'arithmetic.mean' then define it as follows:

arithmetic.mean <- function(x) {
      sum(x)/length(x) }

Notice three things: the calculations are enclosed within curly brackets {}; we don't assign the answer sum(x)/length(x) to a variable name like ybar; and the name of the vector used inside the function (x) may be different from the names on which we might want to use the function in future (like *y*, *w* or *z* for instance). If you type the name of a function on its own, you get a listing of the contents:

arithmetic.mean

```
function(x) {
        sum(x)/length(x) }
```

Now we can test the function on some data. First we use a simple data set where we know the answer already, so that we can check that the function works properly, such as

data < -c(3,4,6,7)

where we can see immediately that the arithmetic mean is 5.

arithmetic.mean(data)

[1] 5

So that's all right. Now we can try it on a realistically big data set

arithmetic.mean(y)

[1] 1.103464

You won't be surprised to learn that R has a built-in function for calculating arithmetic means directly and, again not surprisingly, it is called 'mean'. It works in the same way as our home-made function:

mean(y)

[1] 1.103464

   Arithmetic mean is not the only quantitative measure of central tendency, and in fact it has some rather unfortunate properties. Perhaps the most serious failing of the arithmetic mean is that it is highly sensitive to outliers. Just a single extremely large or extremely small value in the data set will have a big effect on the value of the arithmetic mean. We shall return to this issue later, but our next measure of central tendency does not suffer from being sensitive to outliers. It is called the **median**, and is the 'middle value' in the data set. To write a function to work out the median, the first thing we need to do is sort the data into ascending order:

sorted <- sort(y)

Now we just need to find the middle value. There is a slight difficulty here, because if the vector contains an even number of numbers, then there **is** no middle value. Let's start with the easy case where the vector contains an odd number of numbers. The number of numbers in the vector is given by length(y) and the middle value is half of this:

length(y)/2

[1] 19.5

So the median value is the twentieth value in the sorted data set. To extract the median value of *y* we need to use 20 as a subscript, not 19.5, so we need to convert the value of length(y)/2 into an integer. We use ceiling ('the smallest integer greater than') for this:

ceiling(length(y)/2)

```
[ 1]  20
```

So now we can extract the median value of *y*

sorted[20]

```
[ 1]  1.108847
```

or, more generally

sorted[ceiling(length(y)/2)]

```
[ 1]  1.108847
```

or even more generally, omitting the intermediate variable called sorted:

sort(y)[ceiling(length(y)/2)]

```
[ 1]  1.108847
```

Now what about the case where the vector contains an even number of numbers? Let's manufacture such a vector, by dropping the first element from our vector called *y* using negative subscripts like this:

y.even < -y[-1]
length(y.even)

```
[ 1]  38
```

The logic is that we shall work out the arithmetic average of the two values of *y* on either side of the middle; in this case, the average of the nineteenth and twentieth sorted values:

sort(y.even)[19]

```
[ 1]  1.108847
```

sort(y.even)[20]

```
[ 1]  1.108853
```

So in this case, the median would be

(sort(y.even)[19] + sort(y.even)[20])/2

```
[ 1]  1.108850
```

but to make it general we need to replace the 19 and 20 by length(y.even)/2 and 1 + length(y.even)/2 respectively. The question now arises as to how we know, in general, whether the vector *y* contains an odd or an even number of numbers, so that we can decide which of the two methods to use. The trick here is to use 'modulo'. This is the remainder (the amount 'left over') when one integer is divided by another. An even number has modulo 0 when divided by 2, and an odd number has modulo 1. The modulo function in R is %% (two successive per cent symbols) and it is used where you would use slash (/) to carry out a regular division. You can see this in action with an even number, 38, and odd number, 39:

```
38%%2
```

```
[ 1]  0
```

```
39%%2
```

```
[ 1]  1
```

Now we have all the tools we need to write a general function to calculate medians. Let's call the function med and define it like this:

```
med < -function(x) {
odd.even < -length(x)%%2
if (odd.even = = 0) (sort(x)[length(x)/2] + sort(x)[1 + length(x)/2])/2
else sort(x)[ceiling(length(x)/2)]
}
```

Notice that when the if statement is true (i.e. we have an even number of numbers) then the expression immediately following the if statement is evaluated (this is the code for calculating the median with an even number of numbers). When the if statement is false (i.e. we have an odd number of numbers, and odd.even = = 1) then the expression following the else statement is evaluated (this is the code for calculating the median with an odd number of numbers). Let's try it out, first with the odd-numbered vector *y*, then with the even-numbered vector *y.even*, to check against the values we obtained earlier.

```
med(y)
```

```
[ 1]  1.108847
```

```
med(y.even)
```

```
[ 1]  1.108850
```

Both of these check out. Again, you won't be surprised that there is a built-in function for calculating medians, and helpfully it is called median:

```
median(y)
```

```
[ 1]  1.108847
```

median(y.even)

```
[ 1] 1.108850
```

For processes that change multiplicatively rather than additively, then neither the arithmetic mean nor the median is an ideal measure of central tendency. Under these conditions, the appropriate measure is the **geometric mean**. The formal definition of this is somewhat abstract: the geometric mean is the nth root of the product of the data. If we use capital Greek pi (Π) to represent multiplication, and y 'hat' ($\hat{y}$) to represent the geometric mean, then

$$\hat{y} = \sqrt[n]{\Pi y}.$$

Let's take a simple example we can work out by hand: the numbers of insects on five plants were as follows: 10, 1, 1000, 1, 10. Multiplying the numbers together gives 100 000. There are five numbers, so we want the fifth root of this. Roots are hard to do in your head, so we'll use R as a calculator. Remember that roots are fractional powers, so the fifth root is a number raised to the power $\frac{1}{5} = 0.2$. In R, powers are denoted by the ^ symbol, which is found above the number 6 on the keyboard:

100000^0.2

```
[ 1] 10
```

So the geometric mean of these insect numbers is ten insects per stem. Note that two of the data were exactly like this, so it seems a reasonable estimate of central tendency. The arithmetic mean, on the other hand, is a hopeless measure of central tendency, because the large value (1000) is so influential: $10 + 1 + 1000 + 1 + 10 = 1022$ and $1022/5 = 204.4$. Note that none of the data were close to 204.4, so the arithmetic mean is a poor estimate of central tendency in this case.

insects < -c(1,10,1000,10,1)
mean(insects)

```
[ 1] 204.4
```

Another way to calculate the geometric mean involves the use of logarithms. Recall that to multiply numbers together we add up their logarithms. And to take roots, we divide the logarithm by the root. So we should be able to calculate a geometric mean by finding the antilog (exp) of the average of the logarithms (log) of the data:

exp(mean(log(insects)))

```
[ 1] 10
```

Writing a general function to compute geometric means is one of the exercises.

The use of geometric means draws attention to a general scientific issue. Look at the figure below, which shows numbers varying through time in two populations. Now ask yourself 'which population is the more variable'? The chances are, you will pick the upper line:

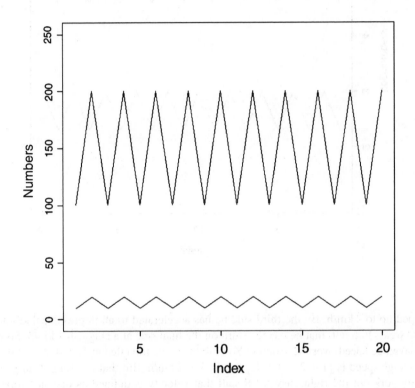

But now look at the scale on the y axis. The upper population is fluctuating 100, 200, 100, 200 and so on. In other words, it is doubling and halving, doubling and halving. The lower curve is fluctuating 10, 20, 10, 20, and so on. It, too, is doubling and halving, doubling and halving. So the answer to the question is: 'they are equally variable'. It is just that one population has a higher mean value than the other (150 vs. 15 in this case). In order not to fall into the trap of saying that the upper curve is more variable than the lower curve, it is good practice to plot the logarithms rather than the raw values of things like population sizes that change multiplicatively.

Now it is clear that both populations are equally variable. Note the change of scale, as specified in the ylim = c(1,6) command (p. 30).

Finally, we should deal with a rather different measure of central tendency. Consider the following problem. An elephant has a territory which is a square of side = 2 km. Each morning, the elephant walks the boundary of this territory. It begins the day at a sedate pace, walking the first side of the territory at a speed of 1 km/h. On the second side, he

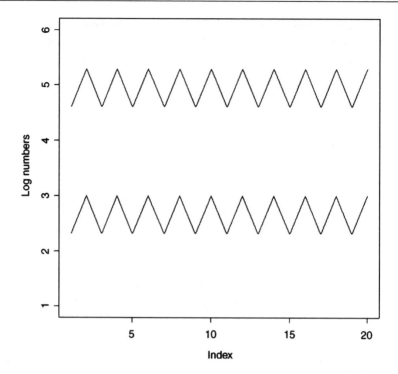

has sped up to 2 km/h. By the third side he has accelerated to an impressive 4 km/h, but this so wears him out, that he has to return on the final side at a sluggish 1 km/h. So what is his average speed over the ground? You might say he travelled at 1, 2, 4 and 1 km/h so the average speed is $(1 + 2 + 4 + 1)/4 = 8/4 = 2$ km/h. But that is wrong. Can you see how to work out the right answer? Recall that velocity is defined as distance travelled divided by time taken. The distance travelled is easy: it's just $4 \times 2 = 8$ km. The time taken is a bit harder. The first edge was 2 km long, and travelling at 1 km/h this must have taken 2 h. The second edge was 2 km long, and travelling at 2 km/h this must have taken 1 h. The third edge was 2 km long and travelling at 4 km/h this must have taken 0.5 h. The final edge was 2 km long and travelling at 1 km/h this must have taken 2 h. So the total time taken was $2 + 1 + 0.5 + 2 = 5.5$ h. So the average speed is not 2 km/h but $8/5.5 = 1.4545$ km/h.

The way to solve this problem is to use the **harmonic mean**. This is the reciprocal of the average of the reciprocals. Remember that a reciprocal is 'one over'. So the average of the speed reciprocals $\frac{1}{1}, \frac{1}{2}, \frac{1}{4}, \frac{1}{1}$ is $2.75 \div 4 = 0.6875$. The reciprocal of this average is the harmonic mean $1/0.6875 = 1.4545$. In symbols, therefore, the harmonic mean, $\tilde{y}$, (y 'curl') is given by

$$\tilde{y} = \frac{1}{\frac{\sum \frac{1}{y}}{n}} = \frac{n}{\sum \frac{1}{y}}.$$

In R, we would write either

```
v < -c(1,2,4,1)
length(v)/sum(1/v)
```

```
[ 1]  1.454545
```

or

```
1/mean(1/v)
```

```
[ 1]  1.454545
```

```
detach(yvals)
rm(v,upper,lower,insects)
```

Getting Help in R

If you know the name of the function that you want to find out about, just type a question mark, ?, followed immediately (without a space) by the name of the function. To find out about graphics parameters (par), for instance, you would type

```
?par
```

If you do not know the exact name of the function, try browsing the index to this book, or use the help.search facility with the name you are looking for in double quotes, like this

```
help.search("read")
```

and a list of the relevant functions involving reading data from files will appear in a window. You can then use ? to look up the function names that look most relevent.

4

Variance

A measure of variability is perhaps the most important quantity in statistical analysis. The greater the variability in the data, the greater will be our uncertainty in the values of parameters estimated from the data, and the lower will be our ability to distinguish between competing hypotheses about the data.

Consider the following data, y, which are plotted simply in the order in which they were measured:

```
y<-c(13,7,5,12,9,15,6,11,9,7,12)
plot(y,ylim=c(0,20))
```

How can we quantify the variation (the scatter) in y that we can see here? Perhaps the simplest measure is the **range** of y values (p. 34):

```
range(y)
```
```
[1]  5   15
```

This is a reasonable measure of variability, but it is too dependent on outlying values for most purposes. Also, we want all of our data to contribute to the measure of variability, not just the maximum and minimum values. How about estimating the mean value, and looking at the departures from the mean (known as 'residuals' or 'deviations')?

The longer these lines, the more variable the data. So this looks promising. How about adding up the lengths of the lines: $\sum (y - \bar{y})$? A moment's thought will show that this is no good, because the negative residuals (from the points below the mean) will cancel out the positive residuals (from the points above the line). In fact, it is easy to prove that this quantity $\sum (y - \bar{y})$ is zero, no matter what the variation in the data, so that's no good (see Box 4.1).

Statistics: An Introduction using R   M. J. Crawley
© 2005 John Wiley & Sons, Ltd   ISBNs: 0-470-02298-1 (PBK); 0-470-02297-3 (PPC)

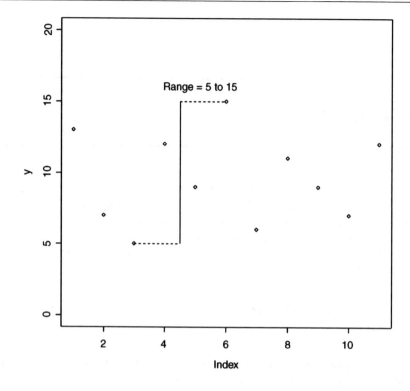

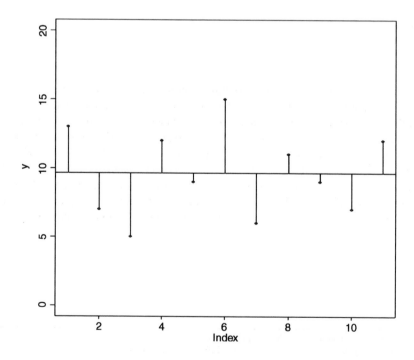

Box 4.1. The sum of the differences $\sum (y - \bar{y})$ is zero

Start by writing down the differences explicitly

$$\sum d = \sum (y - \bar{y}).$$

Take $\sum$ through the brackets. The important point is that $\sum \bar{y}$ is the same as $n.\bar{y}$ so

$$\sum d = \sum y - n\bar{y},$$

and we know that $\bar{y} = \sum y/n$ so

$$\sum d = \sum y - \frac{n \sum y}{n}.$$

The n's cancel, leaving

$$\sum d = \sum y - \sum y = 0.$$

The only problem is the minus signs. How about ignoring the minus signs and adding up the absolute values of the residuals: $\sum (|y - \bar{y}|)$. This is a very good measure of variability, and is used in some modern, computationally intensive methods. The problem is that it makes the sums hard, and we don't want that. A much more straightforward way of getting rid of the problem of the minus signs is to square the residuals before we add them up: $\sum (y - \bar{y})^2$. This is perhaps the most important single quantity in all of statistics. It is called, somewhat unimaginatively, the **sum of squares**. So, in our figure (p. 34) imagine squaring the lengths of each of the vertical lines:

y-mean(y)

```
[ 1] 3.3636364 -2.6363636  -4.6363636  2.3636364 -0.6363636  5.3636364 -3.6363636  1.3636364
[ 9]-0.6363636 -2.6363636   2.3636364
```

(y-mean(y))^2

```
[ 1]11.3140496  6.9504132  21.4958678  5.5867769  0.4049587 28.7685950 13.2231405  1.8595041
[ 9] 0.4049587  6.9504132   5.5867769
```

then adding up all these squared differences:

sum((y-mean(y))^2)

```
[ 1] 102.5455
```

So the sum of squares for our data is 102.5455. But what are its units? Well that depends on the units in which y is measured. Suppose the y values were lengths in mm. So the units of the sum of squares are mm^2 (like an area).

Now what would happen to the sum of squares if we added a twelfth data point? It would get bigger, of course. And it would get bigger for every extra data point we added (except in the unlikely event that our new data point was exactly equal to the mean value, in which case we would add zero squared $= 0$). We don't want our measure of variability to depend on sample size in this way, so the obvious solution is to divide by the number of samples, to get the **mean squared deviation**.

At this point we need to make a brief, but important, diversion. Before we can progress, we need to understand the concept of degrees of freedom.

Degrees of Freedom

Suppose we had a sample of five numbers and their average was 4. What was the sum of the five numbers? It must have been 20, otherwise the mean would not have been 4. So now let's think about each of the five numbers in turn.

We are going to put a number in each of the five boxes. If we allow that the numbers could be positive or negative real numbers, we ask how many values could the first number take. Once you see what I'm doing, you will realize it could take any value. Suppose it was a 2.

2				

How many values could the next number take? It could be anything. Say it was a 7.

2	7			

And the third number could be anything. Suppose it was a 4.

2	7	4		

The fourth number could be anything at all. Say it was 0.

2	7	4	0	

Now, how many values could the last number take? Just one – it **has** to be another 7 because the numbers have to add up to 20 because the mean of the five numbers is 4.

2	7	4	0	7

To recap, we have total freedom in selecting the first number–and the second, third and fourth numbers. However, we have no choice at all in selecting the fifth number. We have four degrees of freedom when we have five numbers. In general we have $(n-1)$ degrees of freedom if we estimated the mean from a sample of size n. More generally still, we can propose a formal definition of degrees of freedom: **degrees of freedom is the sample size, n, minus the number of parameters, p, estimated from the data**. This is so important, you should memorize it. In the example we plotted earlier we had $n = 11$ and we estimated just one parameter from the data, the sample mean, $\bar{y}$. So we have $n - 1 = 10$ d.f.

Variance

We return to developing our quantitative measure of variability. We have come to the conclusion that the sum of squares $\sum (y - \bar{y})^2$ is a good basis for assessing variability, but we have the problem that the sum of squares increases with every new data point we add to the sample. The intuitive thing to do would be to divide by the number of numbers, n, to get the mean squared deviation, but look at the formula for the sum of squares: $\sum (y - \bar{y})^2$. We cannot begin to calculate it until we know the value of the sample mean, $\bar{y}$, and where do we get the value of $\bar{y}$ from? Do we know it in advance? Can we look it up in tables? No, we need to calculate it from the data. The mean value $\bar{y}$ is **a parameter estimated from the data**, so we loose one degree of freedom as a result. Thus, in calculating the mean squared deviation we divide by the degrees of freedom, $n - 1$, rather than by the sample size, n. In the jargon, this provides us with an unbiased estimate of the variance, because we have taken account of the fact that one parameter was estimated from the data prior to computation.

Now we can formalize our definition of the measure that we shall use throughout the book for quantifying variability. It is called **variance** and it is represented conventionally by s^2:

$$\text{variance} = \frac{\text{sum of squares}}{\text{degrees of freedom}}.$$

This is one of the most important definitions in the book, and you should commit it to memory. We can put it into a more mathematical form, by spelling out what we mean by each of the phrases in the numerator and the denominator:

$$\text{variance} = s^2 = \frac{\sum (y - \bar{y})^2}{n - 1}.$$

Let's write an R function to do this. We have most of the necessary components already (see above); the sum of squares is obtained as sum((y-mean(y))^2). For the degrees of freedom, we need to know the number of numbers in the vector, y. This is obtained by the function length(y). Let's call the function variance and write it like this:

```
variance <- function (x) sum((x-mean(x))^2)/(length(x)-1)
```

Now we can try out the function on our data, like this

variance(y)

[1] 10.25455

So there we have it. Our quantification of the variation we saw in the first plot is the sample variance, $s^2 = 10.25455$. You will not be surprised that R provides its own, built-in function for calculating variance, and it has an even simpler name than the function we just wrote: var

var(y)

[1] 10.25455

Variance is used in countless ways in statistical analysis, so this section is probably the most important section in the whole book, and you should re-read it until you are sure that you know exactly what variance is, and precisely what it measures (Box 4.2).

Box 4.2. Short-cut formula for the sum of squares $\sum (y - \bar{y})^2$

The main problem with the formula defining variance is that it involves all those subtractions, $y - \bar{y}$. It would be good to find a simpler way of calculating the sum of squares. Let's expand the bracketed term $(y - \bar{y})^2$ to see if we can make any progress towards a subtraction-free solution:

$$(y - \bar{y})^2 = (y - \bar{y})(y - \bar{y}) = y^2 - 2y\bar{y} + \bar{y}^2.$$

So far, so good. Now we apply the summation

$$\sum y^2 - 2\bar{y} \sum y + n\bar{y}^2 = \sum y^2 - 2\frac{\sum y}{n} \sum y + n\left[\frac{\sum y}{n}\right]^2.$$

Note that only the y's take the summation sign. This is because we can replace $\sum \bar{y}$ by $n\bar{y}$. Now replace $\bar{y}$ with $\sum y/n$ on the right-hand side, then cancel the n's and collect the terms:

$$\sum y^2 - 2\frac{[\sum y]^2}{n} + n\frac{[\sum y]^2}{n^2} = \sum y^2 - \frac{[\sum y]^2}{n}.$$

This is the short-cut formula for computing the sum of squares. It requires only two quantities to be estimated from the data: the sum of the squared y values $\sum y^2$ and the square of the sum of the y values $[\sum y]^2$.

A Worked Example

The data in the following table come from three market gardens. The data show the ozone concentrations in parts per hundred million (pphm) on ten summer days.

```
ozone<-read.table("c:\\temp\\gardens.txt",header=T)
attach(ozone)
ozone
```

	gardenA	gardenB	gardenC
1	3	5	3
2	4	5	3
3	4	6	2
4	3	7	1
5	2	4	10
6	3	4	4
7	1	3	3
8	3	5	11
9	5	6	3
10	2	5	10

The first step in calculating variance is to work out the mean:

```
mean(gardenA)
```

[1] 3

Now we subtract the mean value (3) from each of the data points:

```
gardenA-mean(gardenA)
```

[1] 0 1 1 0 -1 0 -2 0 2 -1

This produces a vector of differences (of length = 10). We need to square these differences:

```
(gardenA-mean(gardenA))^2
```

[1] 0 1 1 0 1 0 4 0 4 1

then add up the squared differences:

```
sum((gardenA-mean(gardenA))^2)
```

[1] 12

This important quantity is called 'the sum of squares'. Variance is the sum of squares divided by degrees of freedom. We have ten numbers, and have estimated one parameter from the data (the mean) in calculating the sum of squares, so we have $10 - 1 = 9$ d.f.

sum((gardenA-mean(gardenA))^2)/9

[1] 1.333333

So the mean ozone concentration in garden A is 3.0 and the variance in ozone concentration is 1.33. We now do the same for garden B:

mean(gardenB)

[1] 5

It has a much higher mean ozone concentration than garden A, but what about its variance?

gardenB-mean(gardenB)

[1] 0   0   1   2 -1 -1 -2   0   1   0

(gardenB-mean(gardenB))^2

[1] 0   0   1   4   1   1   4   0   1   0

sum((gardenB-mean(gardenB))^2)

[1] 12

sum((gardenB-mean(gardenB))^2)/9

[1] 1.333333

This is interesting: although the mean values are quite different, the variances are exactly the same (both have $s^2 = 1.33333$). What about garden C?

mean(gardenC)

[1] 5

Its mean ozone concentration is the same as in garden B.

gardenC-mean(gardenC)

[1] -2 -2 -3 -4   5 -1 -2   6 -2   5

(gardenC-mean(gardenC))^2

[1] 4   4   9   16   25   1   4   36   4   25

sum((gardenC-mean(gardenC))^2)

[1] 128

sum((gardenC-mean(gardenC))^2)/9

[1] 14.22222

So, although the means in gardens B and C are identical, the variances are quite different (1.33 and 14.22 respectively). Are the variances significantly different? We do an *F*-test for this, dividing the larger variance by the smaller variance:

var(gardenC)/var(gardenB)

[1] 10.66667

Then look up the probability of getting an *F*-ratio as big as this by chance alone if the two variances were really the same. We need the cumulative probability of the *F*-distribution, which is a function called **pf** that we need to supply with three **arguments**: the size of the variance ratio (10.667), the number of degrees of freedom in the numerator (9) and the number of degrees of freedom in the denominator (also 9). We did not know in advance which garden was going to have the higher variance, so we do what's called a two-tail test (we simply multiply the probability by 2):

2*(1-pf(10.667,9,9))

[1] 0.001624002

This probability is much less than 5%, so we conclude that there is a highly significant difference between these two variances. We could do this even more simply by using the built-in *F* test:

var.test(gardenB,gardenC)

```
        F test to compare two variances
data:   gardenB and gardenC
F = 0.0938, num df = 9, denom df = 9, p-value = 0.001624
alternative hypothesis: true ratio of variances is not equal to 1
95 percent confidence interval:
 0.02328617 0.37743695
sample estimates:
ratio of variances
          0.09375
```

So the two variances are significantly different, but why does this matter?

What follows is one of the most important lessons so far, so keep re-reading it until you are sure that you understand it. Comparing gardens A and B we can see that two samples can have different means, but the same variance. This is assumed to be the case when we carry out standard tests (like Student's *t*-test) to compare two means, or an analysis of variance to compare three or more means.

Comparing gardens B and C we can see that two samples can have the same mean but different variances. Is it right to say samples with the same mean are identical? No! Let's look into the science in a bit more detail. The damage threshold for lettuces is 8 pphm ozone, so looking at the means shows that both gardens are free of ozone damage on their lettuces (the mean of 5 for both B and C is well below the threshold of 8). Let's look at the raw data for garden B. How many of the days had ozone > 8? Look at the dataframe and you will see that none of the days exceeded the threshold. What about garden C?

```
gardenC
```

```
[ 1] 3   3   2   1 10   4   3 11   3 10
```

In garden C ozone reached damaging concentrations on three days out of ten, so 30% of the time the lettuce plants would be suffering ozone damage. This is the key point: when the variances are different, we should not make inferences by comparing the means. When we compare the means, we conclude that garden C is like garden B, and that there will be no ozone damage to the lettuces. When we look at the data, we see that this is completely wrong: there is ozone damage 30% of the time in garden C and none of the time in garden B.

So, **when the variances are different, don't compare the means**. If you do, you run the risk of coming to entirely the wrong conclusion.

Variance and Sample Size

It is important to understand the relationship between the size of a sample (the replication, *n*) and the value of variance that is estimated. We can do a simple simulation experiment to investigate this:

```
plot(c(0,32),c(0,15),type="n",xlab="Sample size",ylab="Variance")
```

The plan is to select random numbers from a normal distribution using the function rnorm. The distribution is defined as having a mean of 10 and a standard deviation of 2 (this is the square root of the variance, so $s^2 = 4$). We shall work out the variance for sample sizes between $n = 3$ and $n = 31$, and plot 30 independent instances of variance at each of the selected sample sizes:

```
for (df in seq(3,31,2)) {
for( i in 1:30){
x<-rnorm(df,mean=10,sd=2)
points(df,var(x)) }}
```

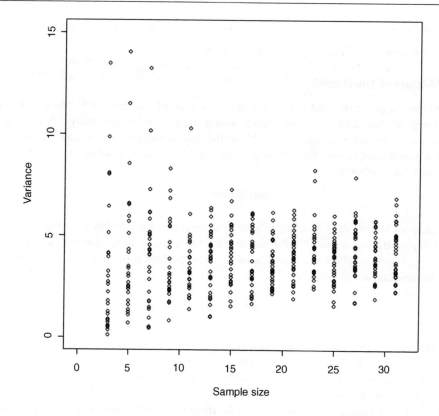

You see that as sample size declines, the range of the estimates of sample variance increases dramatically (remember that the population variance is constant at $s^2 = 4$ throughout). The problem becomes severe below samples of 13 or so, and is very serious for samples of seven or fewer. Even for reasonably large samples (like $n = 31$) the variance varies more than three-fold in just 30 trials (you can see that the rightmost group of points varies from about 2 to about 6). This means that for small samples, the estimated variance is badly behaved, and this has serious consequences for estimation and hypothesis testing.

When people ask 'how many samples do I need?' a statistician will often answer with another question: 'how many can you afford?' Other things being equal, what we have learned in this chapter is that 30 is a reasonably good sample. Anything less than this is a small sample, and anything less than 10 is a very small sample. Anything more than 30 may be an unnecessary luxury (i.e. a waste of resources). We shall see later when we study Power Analysis how the question of sample size can be addressed more objectively, but for the time being take $n = 30$ samples if you can afford it and you won't go far wrong.

Using Variance

Variance is used in two main ways:

- for establishing measures of unreliability (e.g. confidence intervals), and

- for testing hypotheses (e.g. Student's *t*-test).

A Measure of Unreliability

Consider the properties that you would like a measure of unreliability to possess. As the variance of the data increases, what would happen to unreliability of estimated parameters? Would it go up or down? Unreliability would go up as variance increased, so we would want to have the variance on the top (the numerator) of any divisions in our formula for unreliability:

$$\text{unreliability} \propto s^2.$$

What about sample size? Would you want your estimate of unreliability to go up or down as sample size, *n*, increased? You would want unreliability to go down as sample size went up, so you would put sample size on the bottom of the formula for unreliability (i.e. in the denominator):

$$\text{unreliability} \propto \frac{s^2}{n}.$$

Finally, consider the units in which unreliability is measured. What are the units in which our current measure is expressed? Sample size is dimensionless, but variance is based on the sum of squared differences, so it has dimensions of mean squared. So if the mean was a length in cm, the variance would be an area in cm^2. This is an unfortunate state of affairs. It would make good sense to have the dimensions of the unreliability measure and the parameter whose unreliability it is measuring to be the same. That is why all unreliability measures are enclosed inside a big square root term. Unreliability measures are called **standard errors**. What we have just worked out is **the standard error of the mean**

$$SE_{\bar{y}} = \sqrt{\frac{s^2}{n}}.$$

This is a very important equation and should be memorized. Let's calculate the standard errors of each of our market garden means:

```
sqrt(s2A/10)
```
```
[1] 0.3651484
```
```
sqrt(s2B/10)
```
```
[1] 0.3651484
```
```
sqrt(s2C/10)
```
```
[1] 1.19257
```

In written work you should show the unreliability of any estimated parameter in a formal, structured way: 'the mean ozone concentration in Garden A was 3.0 ± 0.365 pphm (1 s.e., $n = 10$)'. You write plus or minus, then the unreliability measure, the units (parts per hundred million in this case) then, in brackets, tell the reader what the unreliability measure is (in this case one standard error) and the size of the sample on which the parameter estimate was based (in this case, 10). This may seem rather stilted, unnecessary even. But the problem is that unless you do this, the reader will not know what kind of unreliability measure you have used. For example, you might have used a 95% confidence interval or a 99% confidence interval instead of one standard error.

Confidence Intervals

A confidence interval shows the likely range in which the mean would fall if the smpling exercise were to be repeated. It is a very important concept that people always find difficult to grasp at first. It is pretty clear that the confidence interval will get wider as the unreliability goes up, so

$$\text{confidence interval} \propto \text{unreliability measure} \propto \sqrt{\frac{s^2}{n}}.$$

But what do we mean by 'confidence'? This is the hard thing to grasp. Ask yourself this question. Would the interval be wider or narrower if we wanted to be **more** confident that our repeat sample mean will fall inside the interval? It may take some thought, but you should be able to convince yourself that the more confident you want to be, the **wider** the interval will need to be. You can see this clearly by considering the limiting case of complete and absolute certainty. Nothing is certain in statistical science, so the interval would have to be infinitely wide.

We can produce confidence intervals of different widths by specifying different levels of confidence. The higher the confidence, the wider the interval. How exactly does this work? How do we turn the proportionality ($\propto$) in the equation above into equality? The answer is by resorting to an appropriate theoretical distribution. Suppose our sample size is too small to use the normal distribution ($n < 30$, as here), then we traditionally use Student's t-distribution. The values of Student's t associated with different levels of confidence are available in the function qt, which gives the quantiles of the t-distribution. Confidence intervals are always two-tailed because the parameter may be larger or smaller than our estimate of it. Thus, if we want to establish a 95% confidence interval we need to calculate the value of Student's t associated with $\alpha = 0.025$ (i.e. with $0.01*(100\%\text{-}95\%)/2$). The value is found like this for the left (0.025) and right-hand (0.975) tails:

```
qt(.025,9)
```

```
[ 1]  -2.262157
```

```
qt(.975,9)
```

```
[ 1]  2.262157
```

The first argument in qt is the probability and the second is the degrees of freedom. This says that values as small as -2.262 standard errors below the mean are to be expected in 2.5% of cases ($p = 0.025$), and values as large as $+2.262$ standard errors above the mean with similar probability ($p = 0.975$). Values of Student's t are **numbers of standard errors** to be expected with specified probability and for a given number of degrees of freedom. The values of t for 99% are bigger than these (0.005 in each tail):

qt(.995,9)

[1] 3.249836

and the value for 99.5% confidence are bigger still (0.0025 in each tail):

qt(.9975,9)

[1] 3.689662

Values of Student's t like these appear in the formula for calculating the width of the confidence interval, and their inclusion is the reason why the width of the confidence interval goes up as our degree of confidence is increased. The other component of the formula, the standard error, is not affected by our choice of confidence level. So, finally, we can write down the formula for the confidence interval of a mean based on a small sample ($n < 30$):

$$\text{confidence interval} = t\text{-value} \times \text{standard error},$$

$$CI_{95\%} = t_{(\alpha=0.025,\text{d.f.}=9)} \sqrt{\frac{s^2}{n}}.$$

For Garden B, therefore, we calculate

qt(.975,9)*sqrt(1.33333/10)

[1] 0.826022

and we would present the result in written work: 'the mean ozone concentration in Garden B was $5.0 \pm 0.826 (95\%$ C.I., $n = 10).$'

Bootstrap

A completely different way of calculating confidence intervals is called bootstrapping. You have probably heard the old phrase about 'pulling yourself up by your own bootlaces'. That is where the term comes from. It is used in the sense of getting 'something for nothing'. The idea is very simple. You have a single sample of n measurements, but you can sample from this in very many ways, so long as you allow some values to appear more than once, and other samples to be left out (i.e. sampling with replacement). All you do is calculate the sample mean lots of times, once for each sampling from your data, then obtain the confidence interval by looking at the extreme highs and lows of the estimated means using a function called **quantile** to extract the interval you want (e.g. a 95% interval is specified using c(0.0275, 0.975) to locate the lower and upper bounds). Here are the data:

```
data<-read.table("c:\\temp\\skewdata.txt",header=T)
attach(data)
names(data)
```

```
[ 1] "values"
```

We shall simulate sample sizes (k) between 5 and 30, and for each sample size we shall take 10 000 independent samples from our data (the vector called values), using the function called **sample** with replacement **(replace=T)**:

```
plot(c(0,30),c(0,60),type="n",xlab="Sample size",ylab="Confidence interval")
for (k in seq(5,30,3)){
a<-numeric(10000)
for (i in 1:10000){
   a[i]<-mean(sample(values,k,replace=T))
}
points(c(k,k),quantile(a,c(.025,.975)),type="b")
}
```

The confidence interval narrows rapidly over the range of sample sizes up to about 20, but more slowly thereafter. At $n = 30$, the bootstrapped CI based on 10 000 simulations was

```
quantile(a,c(.025,.975))
```

```
  2.5%       97.5%
24.86843   37.68985
```

(you will get slightly different values because of the randomization). It is interesting to see how this compares with the Normal theory confidence interval:

$$1.96\sqrt{\frac{s^2}{n}} = 1.96\sqrt{\frac{337.065}{30}} = 6.5698,$$

implying that the sample mean lies in the range 24.39885 to 37.53846. As you see, the estimates from the bootstrap and Normal theory are reassuringly close, but they are not identical.

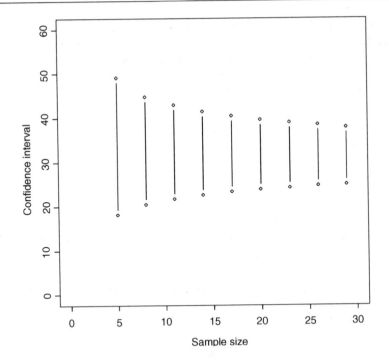

Here are the bootstrapped intervals compared with the intervals calculated from the Normal (solid line) on p. 49:

```
xv<-seq(5,30,0.1)
yv<-mean(values)+1.96*sqrt(var(values)/xv)
lines(xv,yv)
yv<-mean(values)-1.96*sqrt(var(values)/xv)
lines(xv,yv)
```

and Student's t-distribution (dotted line):

```
yv<-mean(values)-qt(.975,xv)*sqrt(var(values)/xv)
lines(xv,yv,lty=2)
yv<-mean(values)+qt(.975,xv)*sqrt(var(values)/xv)
lines(xv,yv,lty=2)
```

For the upper interval, you see that the bootstrapped intervals (vertical lines and open symbols, **type="b"**) fall between the Normal (the lower, solid line) and the Student's t distribution (the greater, dotted line). For the lower interval, however, the bootstrapped intervals are quite different. This is because of the skewness exhibited by these data (see p. 70). Very small values of the response are substantially less likely than predicted by the symmetrical Normal (solid line) or Student's t-distributions (dotted line). Recall that for small-sample confidence intervals using Student's t-distribution, the sample size, n,

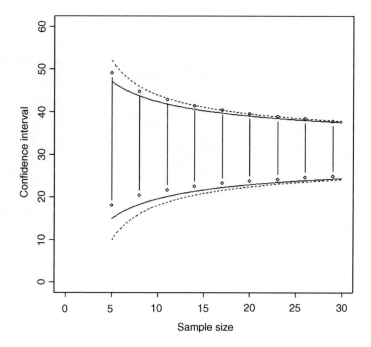

enters the equation twice: once as the denominator in the formula for the standard error of the mean, then again as a determinant of the quantile of the t-distribution $qt(0.975,n)$. That is why the difference between the Normal and the Student's t confidence intervals gets bigger as sample size gets smaller.

So which kind of confidence interval should you choose? I prefer the bootstrapped estimate because it makes fewer assumptions. If, as in our example, the data are skewed, then this is reflected in the asymmetry of the confidence intervals above and below the mean (6.7 above the mean, and 6.1 below it, at $n = 30$). Both Normal and Student's t assume that there is no skew, and so their confidence intervals are symmetrical, whatever the data actually show.

5

Single Samples

Suppose we have a single sample. The questions we might want to answer are these.

- What is the mean value?
- Is the mean value significantly different from current expectation or theory?
- What is the level of uncertainty associated with our estimate of the mean value?

In order to be reasonably confident that our inferences are correct, we need to establish some facts about the distribution of the data.

- Are the values normally distributed or not?
- Are there outliers in the data?
- If data were collected over a period of time, is there evidence for serial correlation?

Non-normality, outliers and serial correlation can all invalidate inferences made by standard parametric tests like Student's t-test. Much better in cases with non-normality and/or outliers to use a non-parametric technique like Wilcoxon's signed-rank test. If there is serial correlation in the data, then you need to use time series analysis or mixed effects models.

Data Summary in the One Sample Case

To see what is involved, read the data called y from the file called das.txt

```
data < -read.table("c:\\temp\\das.txt",header = T)
names(data)
```

```
[ 1]  "y"
```

```
attach(data)
```

Statistics: An Introduction using R   M. J. Crawley
© 2005 John Wiley & Sons, Ltd   ISBNs: 0-470-02298-1 (PBK); 0-470-02297-3 (PPC)

Summarizing the data could not be simpler. We use the built-in function called summary like this:

summary(y)

```
  Min.   1st Qu.   Median    Mean   3rd Qu.    Max.
 1.904    2.241    2.414    2.419    2.568   2.984
```

This gives us six pieces of information about the vector called y. The smallest value is 1.904 (labelled Min. for minimum) and the largest value is 2.984 (labelled Max. for maximum). There are two measures of central tendency: the median is 2.414 and the arithmetic mean in 2.419. What you may be unfamiliar with are the figures labelled '1st Qu.' and '3rd Qu.'. The 'Qu.' is an abbreviation of quartile, which means one quarter of the data: the first quartile is the value of the data, below which lie the smallest 25% of the data. The median is the 2nd quartile by definition (half the data are smaller than the median). The 3rd quartile is the value of the data, above which lie the largest 25% of the data (it is sometimes called the 75th Percentile, because 75% of the values of y are smaller than this value).

Plotting the data requires us to say exactly what sort of plot we want. If we just say

plot(y)

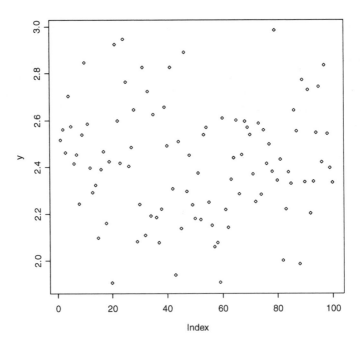

we see the values of y on the y axis, plotted against something called Index. Note that we did not supply the value of Index. Whenever you say plot and there is only one variable,

R assumes that you want to plot the values of *y* in the sequence in which they appear in the dataframe, i.e. starting with the first value on the left, in sequence up to the value in position number = length(y) on the right. This is very useful for data checking to make sure that no really silly values appear in *y* (e.g. typing mistakes on data entry). In the present case, suppose the middle value, y[50] had been typed in as 21.79386 instead of 2.179386. Then plot(y) would draw attention to the mistake at once if you write:

```
y[50] < -21.79386
plot(y)
```

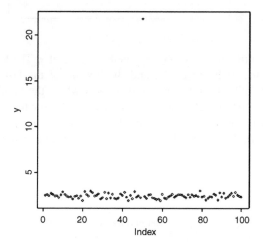

The mistake sticks out like a sore thumb, and the error can be rectified as follows. It is not obvious which of the *y* values is wrong, but it is clear that it is the only value bigger than 10. So to find which value it is, we use the which function like this:

```
which(y > 10)
```

```
[ 1] 50
```

So now we can retype the correct value for the 50th element of *y*:

```
y[50] < - 2.179386
```

and the data are now edited. You could plot them again, to check.

A second kind of plot useful in data summary is the 'box and whisker plot'. It is a visual representation of the data shown in the summary function, above.

```
boxplot(y,ylab = "data values")
```

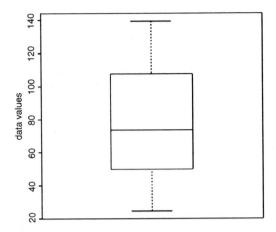

The horizontal bar in the middle shows the median value of y. The top of the box above the median shows the 75th percentile, and the bottom of the box below the median shows the 25th percentile. Both boxes together show where the middle 50% of the data lie (this is called 'the interquartile range'). The whiskers show the maximum and minimum values of y (later on we shall see what happens when the data contain 'outliers').

The last sort of plot that we might want to use for a single sample is the histogram

hist(y)

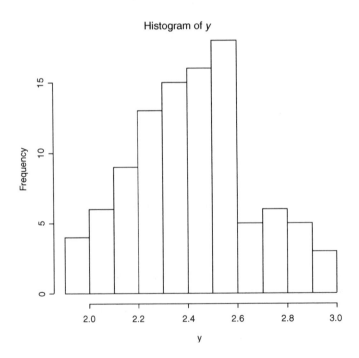

Histogram of y

This is a very informative plot because it shows that the left-hand side of the distribution is a different shape from the right-hand side. The histogram is said to be 'skew to the left' or negatively skew, because there is a longer 'tail' to the left of the mode (where there are six bars) than there is to the right (only four bars).

Simple as they seem, there are actually lots of issues about histograms. Perhaps the most important issue is where, exactly, to draw the lines between the bars (the 'bin widths' in the jargon). For whole number (integer) data this is often an easy decision (draw a bar of the histogram for each of the integer values of y). However, for continuous (real number) data like we have here, that approach is completely inappropriate. How many different values of y do we have in our vector of 100 numbers? The appropriate function to answer questions like this is table: we don't want to see all the values of y, we just want to know how many different values of y there are. That is to say, we want to know the length of the table of different y values:

```
length(table(y))
```

```
[ 1]  100
```

So there are no repeats of any of the y values, and our histogram would be completely uninformative. Let's look more closely to see what R has chosen on our behalf in designing the histogram. The x axis is labelled every 0.2 units, in each of which there are two bars. So the chosen bin width is 0.1. R uses simple rules to select what it thinks will make a 'pretty' histogram. It wants to have a reasonable number of bars (too few bars looks dumpy, while too many makes the shape too rough); there are 11 bars in this case. The next criterion is to have 'sensible' widths for the bins. It makes more sense, for instance to have the bins 0.1 units wide (as here) than to use one tenth of the range of y values, or one eleventh of the range (note the use of the diff and range functions):

```
(max(y)-min(y))/10
```

```
[ 1]  0.1080075
```

```
diff(range(y))/11
```

```
[ 1]  0.09818864
```

So a width of 0.1 is a 'pretty' compromise. As we shall see later, you can specify the width of the bins if you don't like the choice that R has made for you, or if you want to draw two histograms that are exactly comparable.

The Normal Distribution

This famous distribution has a central place in statistical analysis. If you take repeated samples from a population and calculate their averages, then these averages will be normally distributed. This is called the **central limit theorem**. Let's demonstrate it for ourselves. We can take five uniformly distributed random numbers between 0 and 10 and work out the average. The average will be low when we get, say, 2,3,1,2,1 and big when we get 9,8,9,6,8. Typically, of course, the average will be close to 5. Let's do this 10 000 times and look at the distribution of the 10 000 means. The data are rectangularly (uniformly) distributed on the interval 0 to 10, so the distribution of the raw data should be flat-topped:

```
hist(runif(10000)*10,main="")
```

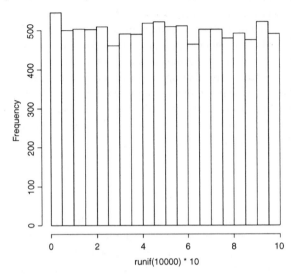

What about the distribution of sample means, based on taking just five uniformly distributed random numbers?

```
means <-numeric(10000)
for (i in 1:10000){
means[i] <- mean(runif(5)*10)
}
hist(means,ylim=c(0,1600))
```

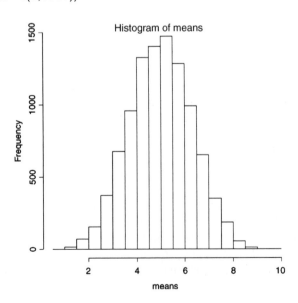

Nice, but how close is this to a Normal distribution? One test is to draw a Normal distribution with the same parameters on top of the histogram but what are these parameters? The Normal is a 'two-parameter distribution' that is characterized by its mean and its standard deviation. We can estimate these two parameters from our sample of 10 000 means (your values will be slightly different because of the randomization):

mean(means)

```
[ 1] 4.998581
```

sd(means)

```
[ 1] 1.289960
```

Now we use these two parameters in the p.d.f. (the 'probability density function') of the normal distribution to create a Normal curve with our particular mean and standard deviation. To draw the smooth line of the Normal curve, we need to generate a series of values for the x axis; inspection of the histograms suggest that sensible limits would be from 0 to 10 (the limits we chose for our uniformly distributed random numbers). A good rule of thumb is that for a smooth curve you need at least 100 values, so let's try this:

xv < -seq(0,10,0.1)

There is just one thing left to do. The p.d.f. has an integral of 1.0 (that's the area beneath the normal curve), but we had 10 000 samples. The scaling factor depends on the total sample size (10 000) and on the bin widths from the histogram (0.5), so the scaling factor $= 10\,000 \times 0.5 = 5000$. To scale the Normal p.d.f. to our particular case, therefore, we multiply by 5000. Finally, we use lines to overlay the smooth curve on our histogram:

yv < -dnorm(xv,mean = 4.998581,sd = 1.28996)*5000
lines(xv,yv)

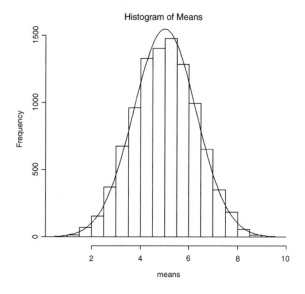

The fit is excellent. The central limit theorem really works. Almost any distribution, even a 'badly behaved' one like the negative binomial (p. 242), will produce a Normal distribution of sample means taken from it.

The great thing about the Normal distribution is that we know so much about its shape. Obviously, all values must lie between − infinity and + infinity, so the area under the whole curve is 1.0. The distribution is symmetrical, so half of our samples will fall below the mean, and half will be above it (i.e. the area beneath the curve to the left of the mean is 0.5). The important thing is that we can predict the distribution of samples in various parts of the curve. For example, about 16% of samples will be more than one standard deviation above the mean, and about 2.5% of samples will be more than two standard deviations below the mean; but how do I know this?

There is an infinite number of different possible Normal distributions: the mean can be anything at all, and so can the standard deviation. For convenience, it is useful to have a standard Normal distribution, whose properties we can tabulate. But what would be a sensible choice for the mean of such a standard normal distribution – obviously not 12.7, but what about 1? Not bad, but the distribution is symmetrical, so it would be good to have the left and right halves with similar scales (not 1 to 4 on the right, but −2 to 1 on the left). The only sensible choice is to have the mean = 0. What about the standard deviation? Should that be 0 as well? Hardly, since that would be a distribution with no spread at all. Not very useful. It could be any positive number, but in practice the most sensible choice is 1. So there you have it. The Standard Normal Distribution is one specific case of the Normal with mean = 0 and standard deviation = 1. So how does this help?

It helps a lot, because now we can work out the area below the curve up to any number of standard deviations (these are the values on the x axis):

```
nd <-seq(-3,3,0.01)
y <-dnorm(nd)
plot(nd,y,type = "l")
```

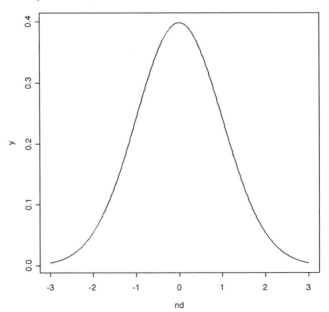

You can see that almost all values fall within three standard deviations of the mean, one way or the other. It is easy to find the area beneath the curve for any value on the x axis (i.e. for any specified value of the standard deviation). Let's start with s.d. $= -2$. What is the area beneath the curve to the left of -2? It is obviously a small number, but the curvature makes it hard to estimate the area accurately from the plot. R provides the answer with a function called pnorm ('probability for a Normal distribution'; strictly 'cumulative probability' as we shall see). Because we are dealing with a standard Normal (mean $= 0$, s.d. $= 1$) we need only specify the value of the Normal deviate, which is -2 in our case:

pnorm(–2)

```
[ 1] 0.02275013
```

This tells us that just a bit less than 2.5% of values will be lower than -2. What about one standard deviation below the mean?

pnorm(–1)

```
[ 1] 0.1586553
```

In this case, about 16% of random samples will be smaller than one standard deviation below the mean. What about big values of the Normal deviate? The histogram shows a maximum of $+3$. What is the probability of getting a sample from a Normal distribution that is more than three standard deviations above the mean? The only point to note here is that pnorm gives the probability of getting a value **less** than the value specified (not more, as we want here). The trick is simply to subtract the value given by pnorm from 1 to get the answer we want:

1-pnorm(3)

```
[ 1] 0.001349898
```

Such a large value is very unlikely indeed – less than a fifth of 1%, in fact.

Probably the most frequent use of the standard Normal distribution is in working out the values of the Normal deviate that can be expected by chance alone. This, if you like, is the opposite kind of problem to the ones we've just been dealing with. There, we provided a value of the Normal deviate (like -1, or -2 or $+3$) and asked what probability was associated with such a value. Now, we want to provide a probability and find out what value of the Normal deviate is associated with that probability. Let's take an important example. Suppose we want to know the upper and lower values of the Normal deviate between which 95% of samples are expected to lie. This means that 5% of samples will lie outside this range, and because the normal is a symmetrical distribution, this means that 2.5% of values will be expected to be smaller than the lower bound (i.e. lie to the left of the lower bound) and 2.5% of values will be expected to be greater than the upper bound (i.e. lie to the right of the lower bound). The function we need is called qnorm ('quantiles of the Normal distribution') and it is used like this by specifying our two probabilities 0.025 and 0.975 in a vector c(0.025,0.975):

```
qnorm(c(0.025,0.975))
```
```
[ 1] -1.959964  1.959964
```

These are two very important numbers in statistics. They tell us that with a Normal distribution, 95% of values will fall between −1.96 and +1.96 standard deviations of the mean. Let's draw these as vertical lines on the normal p.d.f. to see what's involved:

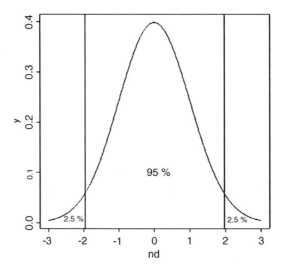

Between the two vertical lines, we can expect 95% of all samples to fall; we expect 2.5% of samples to be less than −1.96 standard deviations below the mean, and we expect 2.5% of samples to be greater than 1.96 standard deviations above the mean. If we discover that this is **not** the case, then our samples are not normally distributed. They might, for instance, follow a Student's t distribution (see p. 67).

To sum up, if we want to provide values of the Normal deviate and work out probabilities, we use pnorm; if we want to provide probabilities and work out values of the Normal deviate, we use qnorm. You should try and remember this important distinction.

Calculations using z of the Normal Distribution

Suppose we have measured the heights of 100 people. The mean height was 170 cm and the standard deviation was 8 cm. The Normal distribution looks like this:

```
ht < -seq(150,190,0.01)
plot(ht,dnorm(ht,170,8),type = "l",ylab = "Probability density",xlab = "Height")
```

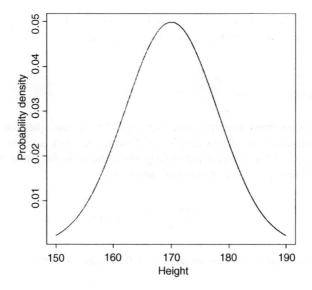

We can ask three sorts of questions about data like these. What is the probability that a randomly selected individual will be:

- shorter than a particular height,
- taller than a particular height,
- between one specified height and another?

The area under the whole curve is exactly 1; everybody has a height between minus infinity and plus infinity. True, but not particularly helpful. Suppose we want to know the probability that one of our people, selected at random from the group, will be less than 160 cm tall. We need to convert this height into a value of z; that is to say, we need to convert 160 cm into a number of standard deviations from the mean. What do we know about the standard Normal distribution? It has a mean of zero and a standard deviation of one. So we can convert any value y, from a distribution with mean $\bar{y}$ and standard deviation s very simply by calculating:

$$z = \frac{(y - \bar{y})}{s}.$$

So we convert 160 cm into a number of standard deviations. It is less than the mean height (170 cm) so its value will be negative:

$$z = \frac{(160 - 170)}{8} = -1.25.$$

Now we need to find the probability of a value of the standard normal taking a value of -1.25 or smaller. This is the area under the left-hand tail of the distribution. The function we need for this is pnorm: we provide it with a value of z (or, more generally, with a quantile) and it provides us with the probability we want:

pnorm(-1.25)

[1] 0.1056498

So the answer to our first question is just over 10%. The second question is: what is the probability of selecting one of our people and finding that they are taller than 185 cm? The first two parts of the exercise are exactly the same as before; first we convert our value of 185 cm into a number of standard deviations:

$$z = \frac{(185 - 170)}{8} = 1.875;$$

then we ask what probability is associated with this, using pnorm:

pnorm(1.875)

[1] 0.9696036

But this is the answer to a different question. This is the probability that someone will be **less** than 185 cm tall (that is what the function pnorm has been written to provide). All we need to do is to work out the complement of this:

1-pnorm(1.875)

[1] 0.03039636

So the answer to the second question is about 3%. Finally, we might want to know the probability of selecting a person between 165 cm and 180 cm? We have a bit more work to do here, because we need to calculate two z values:

$$z_1 = \frac{(165 - 170)}{8} = -0.625 \text{ and } z_2 = \frac{(180 - 170)}{8} = 1.25.$$

The important point to grasp is this: we want the probability of selecting a person between these two z values, so we subtract the smaller probability from the larger probability. It might help to sketch the Normal curve and shade in the area you are interested in:

pnorm(1.25)-pnorm(-0.625)

[1] 0.6283647

Thus we have a 63% chance of selecting a medium sized person (taller than 165 cm and shorter than 180 cm) from this sample with a mean height of 170 cm and a standard deviation of 8 cm.

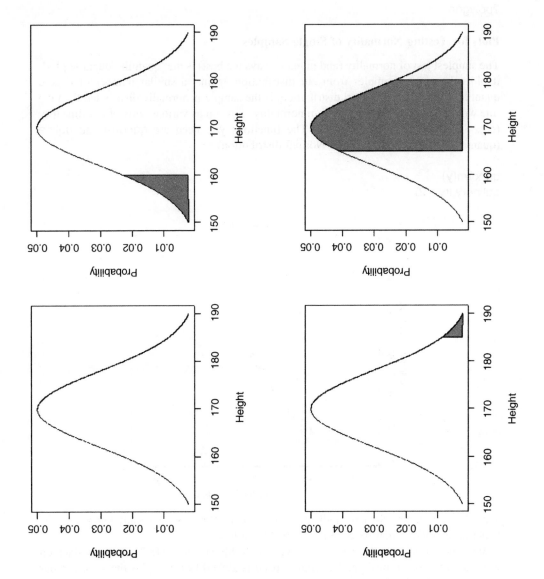

The function called polygon is used for colouring-in different shaped areas under the curve: to see how it is used refer to the figure-generating code on the web site (see Preface), or type

?polygon

Plots for Testing Normality of Single Samples

The simplest test of normality (and in many ways the best) is the 'quantile–quantile plot'; it plots the ranked samples from our distribution against a similar number of ranked quantiles taken from a normal distribution. If the sample is normally distributed then the line will be straight. Departures from normality show up as various sorts of non-linearity (e.g. S-shapes, or banana shapes). The functions you need are qqnorm and qqline (quantile–quantile plot against a Normal distribution):

qqnorm(y)
qqline(y,lty = 2)

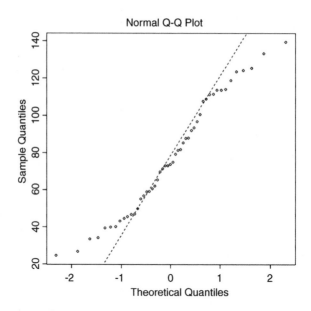

This shows a marked S-shape, indicative of non-normality (as we already know, our distribution is non-Normal because it is skew to the left).

We can investigate the issues involved with Michelson's (1879) famous data on estimating the speed of light. The actual speed is $299\,000\ \mathrm{km\ s^{-1}}$ plus the values in our dataframe called light:

light < -read.table("c:\\temp\\light.txt",header = T)
attach(light)
names(light)

[1] "speed"

hist(speed)

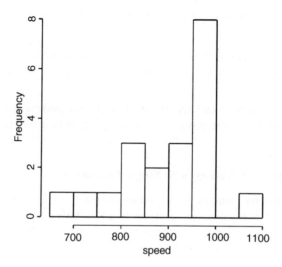

We get a **summary** of the non-parametric descriptors of the sample like this:

summary(speed)

```
Min.    1st Qu.  Median    Mean   3rd Qu.   Max.
650         850     940     909       980   1070
```

From this, you see at once that the median (940) is substantially bigger than the mean (909), as a consequence of the strong negative skew in the data seen in the histogram. The **interquartile range** is the difference between the 1st and 3rd quartiles: $980 - 850 = 130$. This is useful in the detection of outliers: a good rule of thumb is that an **outlier** is a value more than 1.5 times the interquartile range above the 3rd quartile, or below the 1st quartile ($130 \times 1.5 = 195$). In this case, therefore, outliers would be measurements of speed that were less than $850 - 195 = 655$ or greater than $980 + 195 = 1175$. You will see that there are no large outliers in this data set, but one or more small outliers (the minimum is 650).

Inference in the One-sample Case

We now know that the speed of light is 299 792.458 km/s. We want to test the hypothesis that Michelson's estimate of the speed of light is significantly different from the value of 299 990 km/s thought to prevail at the time. The data have all had 299 000 subtracted from them, so the test value is 990. Because of the non-Normality, the use of Student's *t*-test in this case is ill advised. The correct test is Wilcoxon's signed rank test. The code for this is in a library of 'Classical Tests' called ctest:

wilcox.test(speed,mu = 990)

```
    Wilcoxon signed rank test with continuity correction
data: speed
V = 22.5, p-value = 0.00213
alternative hypothesis: true mu is not equal to 990
```

```
Warning message:
Cannot compute exact p-value with ties in: wilcox.
test.default (speed, mu = 990)
```

We reject the null hypothesis and accept the alternative hypothesis because $p = 0.00213$ (i.e. much less than 0.05). The speed of light is significantly less than 990.

Bootstrap in Hypothesis Testing with Single Samples

We shall meet parametric methods for hypothesis testing later. Here we use bootstrapping to illustrate another non-parametric method of hypothesis testing. Our sample mean value of y is 909. The question we have been asked to address is this: 'how likely is it that the population mean that we are trying to estimate with our random sample of 100 values is as big as 990?'.

We take 10 000 random samples with replacement using $n = 100$ from the 100 values of light and calculate 10 000 values of the mean. Then we ask: what is the probability of obtaining a mean as large as 990 by inspecting the right-hand tail of the cumulative probability distribution of our 10 000 bootstrapped mean values? This is not as hard as it sounds:

```
a < -numeric(10000)
for(i in 1:10000) a[i] < -mean(sample(speed,replace = T))
hist(a)
```

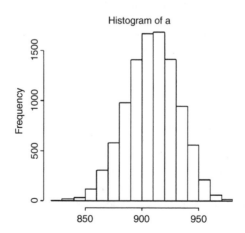

The test value of 990 is off the scale to the right. A mean of 990 is clearly most unlikely, given the data:

```
max(a)
```

```
[ 1]  979
```

In our 10 000 samples of the data, we never obtained a mean value greater than 979, so the probability that the mean is 990 is clearly $p < 0.0001$.

Student's *t*-distribution

Student's *t*-distribution is used instead of the Normal distribution when sample sizes are small ($n < 30$). Remember that the 95% intervals of the standard normal were -1.96 to $+1.96$ standard deviations. Student's *t*-distribution produces bigger intervals than this. The smaller the sample, the bigger the interval. Let's see this in action. The equivalents of pnorm and qnorm are pt and qt. We are going to plot a graph to show how the upper interval (equivalent to the Normal's 1.96) varies with sample size in a *t*-distribution. This is a deviate so the appropriate function is qt. We need to supply it with the probability (in this case $p = 0.975$) and the degrees of freedom (we'll vary these from 1 to 30 to produce the graph)

```
plot(c(0,30),c(0,10),type = "n",xlab = "Degrees of freedom",ylab = "Students t value")
lines(1:30,qt(0.975,df = 1:30))
```

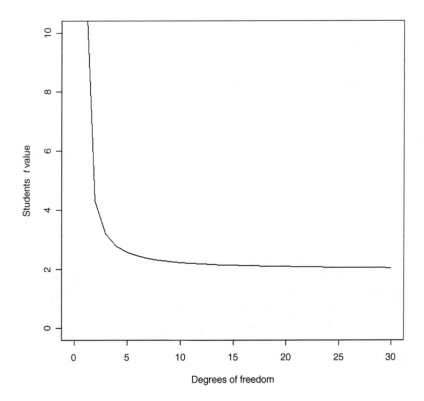

The importance of using Student's *t* rather than the Normal is relatively slight until the degrees of freedom fall below about ten (above which the critical value is roughly two), and then it increases dramatically below about five degrees of freedom. For samples with more than 30 degrees of freedom, Student's *t* produces an asymptotic value of 1.96, just like the Normal. This graph demonstrates that Student's $t = 2$ is a reasonable rule of thumb; memorizing this will save you lots of time in looking up critical values in later life.

So what does the *t*-distribution look like, compared to a Normal distribution? Let's redraw the standard normal as a dotted line (lty = 2):

```
xvs <-seq(-4,4,0.01)
plot(xvs,dnorm(xvs),type = "l",lty = 2,ylab = "Probability density", xlab = "Deviates")
```

Now we can overlay Student's *t* with d.f. = 5 as a solid line to see the difference:

```
lines(xvs,dt(xvs,df = 5))
```

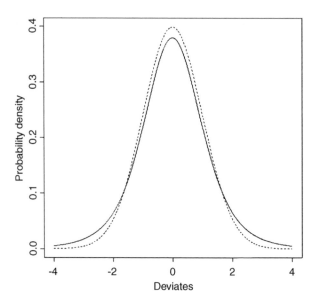

The difference between the Normal (dotted) and Student's *t*-distributions (solid line) is that the *t*-distribution has 'fatter tails'. This means that extreme values are more likely with a *t*-distribution than with a Normal, and the confidence intervals are correspondingly broader. So instead of a 95% interval of ±1.96 with a Normal distribution we should have a 95% interval of ±2.57 for a Student's *t*-distribution with five degrees of freedom:

```
qt(0.975,5)
```
```
[ 1] 2.570582
```

Higher-order Moments of a Distribution

So far, and without saying so explicitly, we have encountered the first two moments of a sample distribution. The quantity $\sum y$ was used in the context of defining the arithmetic mean of a single sample: this is the first moment $\bar{y} = \sum y/n$. The quantity $\sum (y - \bar{y})^2$, the sum of squares, was used in calculating sample variance, and this is the second moment of the distribution, $s^2 = \sum (y - \bar{y})^2/(n - 1)$. Higher-order moments involve powers of the difference greater than two like $\sum (y - \bar{y})^3$ and $\sum (y - \bar{y})^4$.

Skew

Skew (or skewness) is the dimensionless version of the third moment about the mean

$$m_3 = \frac{\sum (y - \bar{y})^3}{n}$$

which is rendered dimensionless by dividing by the cube of the standard deviation of y (because this is also measured in units of y^3):

$$s_3 = \text{s.d.} (y)^3 = (\sqrt{s^2})^3.$$

The skew is then given by

$$\text{skew} = \gamma_1 = \frac{m_3}{s_3}.$$

It measures the extent to which a distribution has long, drawn out **tails** on one side or the other. A Normal distribution is symmetrical and has skew $= 0$. Negative values of γ_1 mean skew to the left (negative skew) and positive values mean skew to the right. To test whether a particular value of skew is significantly different from 0 (and hence the distribution from which it was calculated is significantly non-Normal) we divide the estimate of skew by its approximate standard error:

$$\text{s.e.}_{\gamma_1} = \sqrt{\frac{6}{n}}.$$

It is straightforward to write an R function to calculate the degree of skew for any vector of numbers, x, like this:

```
skew <-function(x){
m3 <-sum((x-mean(x))^3)/length(x)
s3 <-sqrt(var(x))^3
m3/s3 }
```

Note the use of the length(x) function to work out the sample size, n, whatever the size of the vector x. The last expression inside a function is not assigned to a variable name, and is returned as the value of skew(x) when this is executed from the command line.

```
data <-read.table("c:\\temp\\skewdata.txt",header=T)
attach(data)
names(data)
```

```
[1] "values"
```

```
hist(values)
```

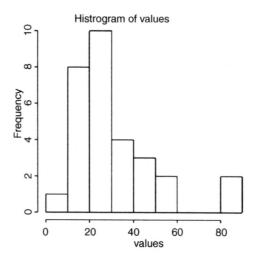

Histrogram of values

The data appear to be positively skewed (i.e. to have a longer tail on the right than on the left). We use the new function skew to quantify the degree of skewness:

```
skew(values)
```

```
[1] 1.318905
```

Now we need to know whether a skew of 1.319 is significantly different from zero. We do a t-test, dividing the observed value of skew by its standard error $\sqrt{6/n}$

```
skew(values)/sqrt(6/length(values))
```

```
[1] 2.949161
```

Finally we ask, what is the probability of getting a t-value of 2.949 by chance alone, when the skew value really is zero?

```
1-pt(2.949,28)
```

```
[1] 0.003185136
```

We conclude that these data show significant non-Normality ($p < 0.0032$).

The next step might be to look for a transformation that normalizes the data by reducing the skewness. One way of drawing in the larger values is to take square roots, so let's try this to begin with:

skew(sqrt(values))/sqrt(6/length(values))

[1] 1.474851

This is not significantly skew. Alternatively, we might take the logs of the values:

skew(log(values))/sqrt(6/length(values))

[1] -0.6600605

This is now slightly skew to the left (negative skew), but the value of Student's t is smaller than with a square root transformation, so we might prefer a log transformation in this case.

Kurtosis

This is a measure of non-Normality that has to do with the peakyness, or flat-toppedness, of a distribution. The Normal distribution is bell shaped, whereas a kurtotic distribution is other than bell shaped. In particular, a more flat-topped distribution is said to be platykurtotic, and a more pointy distribution is said to be leptokurtotic. Kurtosis is the dimensionless version of the fourth moment about the mean

$$m_4 = \frac{\sum (y - \bar{y})^4}{n},$$

which is rendered dimensionless by dividing by the square of the variance of y (because this is also measured in units of y^4):

$$s_4 = \text{var}(y)^2 = (s^2)^2.$$

Kurtosis is then given by

$$\text{kurtosis} = \gamma_2 = \frac{m_4}{s_4} - 3.$$

The minus 3 is included because a Normal distribution has $m_4/s_4 = 3$. This formulation therefore has the desirable property of giving zero kurtosis for a Normal distribution, while a flat-topped (platykurtic) distribution has a negative value of kurtosis, and a pointy (leptokurtic) distribution has a positive value of kurtosis. The approximate standard error of kurtosis is

$$\text{s.e.}_{\gamma_2} = \sqrt{\frac{24}{n}}.$$

An R function to calculate kurtosis might look like this:

```
kurtosis < -function(x) {
m4 < -sum((x-mean(x))^4)/length(x)
s4 < -var(x)^2
m4/s4 - 3 }
```

For our present data, we find that kurtosis is not significantly different from Normal:

```
kurtosis(values)
```

```
[ 1] 1.297751
```

```
kurtosis(values)/sqrt(24/length(values))
```

```
[ 1] 1.450930
```

6

Two Samples

There is absolutely no point in carrying out an analysis that is more complicated than it needs to be. Occam's razor applies to the choice of statistical model just as strongly as to anything else: simplest is best. The so-called classical tests deal with some of the most frequently-used kinds of analysis, and they are the models of choice for:

- comparing two variances (Fisher's F test, var.test),

- comparing two sample means with normal errors (Student's t-test, t.test),

- comparing two sample means with non-normal errors (Wilcoxon's rank test, wilcox.test),

- comparing two proportions (the binomial test, prop.test),

- correlating two variables (Pearson's or Spearman's rank correlation, cor.test),

- testing for independence in contingency tables (chi-square test, chisq.test or Fisher's exact test, fisher.test).

Comparing Two Variances

Before we can carry out a test to compare two sample means, we need to test whether the sample variances are significantly different (see p. 42). The test could not be simpler. It is called Fisher's F test after the famous statistician and geneticist R. A. Fisher, who worked at Rothamsted, UK. To compare two variances, all you do is **divide the larger variance by the smaller variance**.

Obviously, if the variances are the same, the ratio will be 1. In order to be significantly different, the ratio will need to be significantly bigger than 1 (because the larger variance goes on top, in the numerator). How will we know a significant value of the variance ratio from a non-significant one? The answer, as always, is to look up the **critical value** of the variance ratio. In this case, we want critical values of Fisher's F. The R function for this is qf which stands for 'quantiles of the F distribution'. For our example of ozone levels in market gardens (see p. 39) there were ten replicates in each garden, so there were $10 - 1 = 9$ degrees of freedom for each garden. In comparing two gardens, therefore, we have 9 d.f. in the numerator and 9 d.f. in the denominator. Although F tests in analysis of

Statistics: An Introduction using R   M. J. Crawley
© 2005 John Wiley & Sons, Ltd   ISBNs: 0-470-02298-1 (PBK); 0-470-02297-3 (PPC)

variance are typically one-tailed (the treatment variance is expected to be larger than the error variance if the means are significantly different, see p. 41), in this case, we had no expectation as to which garden was likely to have the higher variance, so we carry out a two-tailed test ($p = 1 - \alpha/2$). Suppose we work at the traditional $\alpha = 0.05$, then we find the critical value of F like this:

```
qf(0.975,9,9)
```

```
4.025994
```

This means that a calculated variance ratio will need to be greater than or equal to 4.02 in order for us to conclude that the two variances are significantly different at $\alpha = 0.05$. To see the test in action, we can compare the variances in ozone concentration for market gardens B and C:

```
f.test.data < -read.table("c:\\temp\\f.test.data.txt",header = T)
attach(f.test.data)
names(f.test.data)
```

```
[ 1] "gardenB" "gardenC"
```

First, we compute the two variances:

```
var(gardenB)
```

```
[ 1] 1.333333
```

```
var(gardenC)
```

```
[ 1] 14.22222
```

The larger variance is clearly in garden C, so we compute the F ratio like this:

```
F.ratio < -var(gardenC)/var(gardenB)
F.ratio
```

```
[ 1] 10.66667
```

The variance in garden C is more than ten times as big as the variance in garden B. The critical value of F for this test (with 9 d.f. in both the numerator and the denominator) is 4.026 (see qf, above), so we conclude that **since the calculated value is larger than the critical value we reject the null hypothesis**. The null hypothesis was that the two variances were not significantly different, so we accept the alternative hypothesis that the two variances are significantly different. In fact, it is better practice to present the p value associated with the calculated F ratio rather than just to reject the null hypothesis; to do this we use pf rather than qf. We double the resulting probability to allow for the two-tailed nature of the test:

```
2*(1-pf(F.ratio,9,9))
```

```
[ 1] 0.001624199
```

so the probability that the variances are the same is $p < 0.002$. Because the variances are significantly different, it would be wrong to compare the two sample means using Student's t-test.

There is a built-in function called var.test for speeding up the procedure. All we provide are the names of the two variables containing the raw data whose variances are to be compared (we don't need to work out the variances first):

var.test(gardenB,gardenC)

```
        F test to compare two variances

data: gardenB and gardenC
F = 0.0938, num df = 9, denom df = 9, p-value = 0.001624
alternative hypothesis: true ratio of variances is not equal to 1
95 percent confidence interval:
  0.02328617 0.37743695
sample estimates:
ratio of variances
          0.09375
```

Note that the variance ratio, F, is given as roughly $\frac{1}{10}$ rather than roughly 10 because var.test put the variable name that came first in the alphabet (garden B) on top (i.e. in the numerator) instead of the bigger of the two variances. However, the p value of 0.0016 is correct, and we reject the null hypothesis. These two variances are highly significantly different.

Comparing Two Means

The question is this: given what we know about the variation from replicate to replicate within each sample (the within-sample variance), how likely is it that our two sample means were drawn from populations with the same average? If the answer is highly likely, then we shall say that our two sample means are not significantly different. If it is rather unlikely, then we shall say that our sample means are significantly different. Perhaps a better way to proceed is to work out the probability that the two samples were indeed drawn from populations with the same mean. If this probability is very low (say, less than 5% or less than 1%) then we can be reasonably certain (95% or 99% in these two examples) that the means really are different from one another. Note, however, that we can never be 100% certain; the apparent difference might just be due to random sampling – we just happened to get a lot of low values in one sample, and a lot of high values in the other.

There are two simple tests for comparing two sample means:

- **Student's t-test** when the samples are independent, the variances constant, and the errors are Normally distributed, or

- **Wilcoxon rank sum test** when the samples are independent but the errors are **not** Normally distributed (e.g. they are ranks or scores of some sort).

What you should do when these assumptions are violated (e.g. when the variances are different) is discussed later on.

Student's t-Test

Student was the pseudonym of W.S. Gosset who published his influential paper in *Biometrika* in 1908. He was prevented from publishing under his own name by dint of the archaic employment laws in place at the time, which allowed his employer, the Guinness Brewing Company, to prevent him publishing independent work. Student's t-distribution, later perfected by R. A. Fisher, revolutionized the study of small sample statistics where inferences need to be made on the basis of the sample variance s^2 with the population variance σ^2 unknown (indeed, usually unknowable). The test statistic is the number of standard errors by which the two sample means are separated:

$$t = \frac{\text{difference between the two means}}{\text{s.e. of the difference}} = \frac{\bar{y}_A - \bar{y}_B}{\text{s.e.}_{\text{diff}}}.$$

Now we know the standard error of the mean (see p. 44) but we have not yet met the standard error of the difference between two means. For two independent (i.e. non-correlated) variables, **the variance of a difference is the sum of the separate variances** (see Box 6.1).

Box 6.1. The variance of a difference between two independent samples

We want to work out the sum of squares of a difference between samples A and B. First we express each y variable as a departure from its own mean, μ

$$\sum [(y_A - \mu_A) - (y_B - \mu_B)]^2.$$

If we were to divide by the degrees of freedom, we would get the variance of the difference, $\sigma^2_{\bar{y}_A - \bar{y}_B}$. Start by calculating the square of the difference:

$$(y_A - \mu_A)^2 + (y_B - \mu_B)^2 - 2(y_A - \mu_A)(y_B - \mu_B),$$

then apply summation

$$\sum (y_A - \mu_A)^2 + \sum (y_B - \mu_B)^2 - 2 \sum (y_A - \mu_A)(y_B - \mu_B).$$

We already know that the average of $\sum (y_A - \mu_A)^2$ is the variance of population A and the average of $\sum (y_B - \mu_B)^2$ is the variance of population B (see Box 4.2). So the variance of the **difference** between the two sample means is the **sum** of the variances of the two samples, minus a term $= 2 \sum (y_A - \mu_A)(y_B - \mu_B)$, i.e. minus two times the covariance of samples A and B (see Box 6.2). However, because the samples from A and B are independently drawn they are uncorrelated, the covariance is zero, and so $2\sum(y_A - \mu_A)(y_B - \mu_B) = 0$. This important result needs to be stated separately

$$\sigma^2_{\bar{y}_A - \bar{y}_B} = \sigma^2_A + \sigma^2_B.$$

So if two samples are independent, **the variance of the difference is the sum of the two sample variances**. This is **not** true, of course, if the samples are positively or negatively correlated (see p. 97).

This important result allows us to write down the formula for the **standard error of the difference** between two sample means

$$\text{s.e.}_{\text{difference}} = \sqrt{\frac{s_A^2}{n_A} + \frac{s_B^2}{n_B}}.$$

At this stage we have everything we need to carry out Student's t-test. Our null hypothesis is that the two sample means are the same, and we shall accept this unless the value of Student's t is so large that it is unlikely that such a difference could have arisen by chance alone. For the ozone example introduced on p. 39, each sample has nine degrees of freedom, so we have 18 d.f. in total. Another way of thinking of this is to reason that the complete sample size is 20, and we have estimated two parameters from the data, $\bar{y}_A$ and $\bar{y}_B$, so we have $20 - 2 = 18$ d.f. We typically use 5% as the chance of rejecting the null hypothesis when it is true (this is the Type I error rate). Since we didn't know in advance which of the two gardens was going to have the higher mean ozone concentration (and we usually don't), this is a two-tailed test, so the **critical value** of Student's t is:

```
qt(0.975,18)
```

```
[1] 2.100922
```

This means that our test statistic needs to be bigger than 2.1 in order to reject the null hypothesis, and hence to conclude that the two means are significantly different at $\alpha = 0.05$. The dataframe is attached like this:

```
t.test.data < -read.table("c:\\temp\\t.test.data.txt",header=T)
attach(t.test.data)
names(t.test.data)
```

```
[1] "gardenA" "gardenB"
```

A useful graphical test for two samples employs the 'notches' option of boxplot:

```
ozone < -c(gardenA,gardenB)
label < -factor(c(rep("A",10),rep("B",10)))
boxplot(ozone~label,notch=T,xlab="Garden",ylab="Ozone")
```

Because the notches of two plots do not overlap, we conclude that the medians are significantly different at the 5% level. Note that the variability is similar in both gardens (both in terms of the range – the whiskers – and the inter-quartile range – the boxes).

To carry out a t-test longhand, we begin by calculating the variances of the two samples, s2A and s2B:

```
s2A < -var(gardenA)
s2B < -var(gardenB)
```

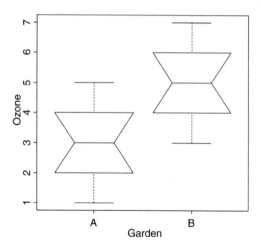

The value of the test statistic for Student's t is: **the difference divided by the standard error of the difference**. The numerator is the difference between the two means, and the denominator is the square root of the sum of the two variances divided by their sample sizes:

(mean(gardenA)-mean(gardenB))/sqrt(s2A/10+s2B/10)

which gives the value of Student's t as

```
[ 1] -3.872983
```

With t-tests you can ignore the minus sign; it is only the absolute value of the difference between the two sample means that concerns us. So the calculated value of the test statistic is 3.87 and the critical value is 2.10 (qt(0.975,18), above). This can be written: **since the calculated value is larger than the critical value we reject the null hypothesis**. Notice that the wording is exactly the same as it was for the F test (above). Indeed, the wording is always the same for all kinds of tests, and you should try to memorize it. The abbreviated form is easier to remember: **larger reject, smaller accept**. The null hypothesis was that the two means are not significantly different, so we reject this and accept the alternative hypothesis that the two means are significantly different. Again, rather than merely rejecting the null hypothesis, it is better to state the probability that data as extreme as this (or more extreme) would be observed if the mean values were the same. For this we use pt rather than qt, and $2 \times$ pt because we are doing a two-tailed test:

2*pt(-3.872983,18)

```
[ 1]  0.001114540
```

so $p < 0.0015$. You won't be surprised to learn that there is a built-in function to do all the work for us. It is called, helpfully, t.test and is used simply by providing the names of the two vectors containing the samples on which the test is to be carried out (garden A and garden B in our case):

t.test(gardenA,gardenB)

There is rather a lot of output. You often find this – the simpler the statistical test, the more voluminous the output.

```
      Welch Two Sample t-test
data: gardenA and gardenB
t = -3.873, df = 18, p-value = 0.001115
alternative hypothesis: true difference in means is not equal to 0
95 percent confidence interval:
 -3.0849115 -0.9150885
sample estimates:
mean of x mean of y
        3         5
```

The result is exactly the same as we obtained longhand. The value of t is -3.873 and since the sign is irrelevant in a t test we reject the null hypothesis because the test statistic is larger than the critical value of 2.1. The mean ozone concentration is significantly higher in garden B than in garden A. The computer print-out also gives a p value and a confidence interval. Note that, because the means are significantly different, the confidence interval on the difference does not include zero (in fact, it goes from -3.085 up to -0.915). You might present the result like this: ozone concentration was significantly higher in garden B (mean $= 5.0$ p.p.h.m.) than in garden A (mean $= 3.0$ p.p.h.m.; $t = 3.873$, $p = 0.0011$ (two-tailed), d.f. $= 18$).

Wilcoxon Rank Sum Test

This is a non-parametric alternative to Student's t-test, which we could use if the errors were non-Normal. The Wilcoxon rank sum test statistic, W, is calculated as follows. Both samples are put into a single array with their sample names clearly attached (A and B in this case, as explained below). Then the aggregate list is sorted, taking care to keep the sample labels with their respective values. A rank is assigned to each value, with ties getting the appropriate average rank (two-way ties get (rank $i +$ (rank $i + 1$))/2, three-way ties get (rank $i +$ (rank $i + 1$) + (rank $i + 3$))/3, and so on). Finally the ranks are added up for each of the two samples, and significance is assessed on size of the smaller sum of ranks.

First we make a combined vector of the samples

ozone < -c(gardenA,gardenB)
ozone

[1] 3 4 4 3 2 3 1 3 5 2 5 5 6 7 4 4 3 5 6 5

then make a list of the sample names, A and B

```
label < -c(rep("A",10),rep("B",10))
label
```

```
[ 1] "A" "A" "A" "A" "A" "A" "A" "A" "A" "A" "B" "B" "B" "B" "B" "B" "B" "B" "B" "B"
```

Now use the built-in function rank to get a vector containing the ranks, smallest to largest, within the combined vector:

```
combined.ranks < -rank(ozone)
combined.ranks
```

```
[ 1]    6.0 10.5 10.5  6.0   2.5 6.0 1.0 6.0 15.0 2.5  15.0 15.0 18.5 20.0 10.5
[ 16] 10.5  6.0 15.0 18.5 15.0
```

Notice that the ties have been dealt with by averaging the appropriate ranks. Now all we need to do is calculate the sum of the ranks for each garden. We use tapply with sum as the required operation

```
tapply(combined.ranks,label,sum)
```

```
  A    B
 66  144
```

Finally, we compare the smaller of the two values (66) with values in Tables of Wilcoxon rank sums (e.g. Snedecor and Cochran 1980: p. 555), and reject the null hypothesis if our value of 66 is **smaller** than the value in tables. For samples of size ten and ten like ours, the 5% value in tables is 78. Our value is smaller than this, so we reject the null hypothesis. The two sample means are significantly different (in agreement with our earlier *t*-test).

We can carry out the whole procedure automatically, and avoid the need to use tables of critical values of Wilcoxon rank sums, by using the built-in function wilcox.test:

```
wilcox.test(gardenA,gardenB)
```

which produces the following output:

```
        Wilcoxon rank sum test with continuity correction
data: gardenA and gardenB
W = 11, p-value = 0.002988
alternative hypothesis: true mu is not equal to 0

Warning message:
Cannot compute exact p-value with ties in:
wilcox.test.default(gardenA, gardenB)
```

The function uses a normal approximation algorithm to work out a z value, and from this a p value to assess the hypothesis that the two means are the same. This p value of 0.002988 is much less than 0.05, so we reject the null hypothesis, and conclude that the mean ozone concentrations in gardens A and B are significantly different. The warning message at the end draws attention to the fact that there are ties in the data (repeats of the same ozone measurement), and this means that the p value cannot be calculated exactly (this is seldom a real worry).

It is interesting to compare the p values of the t-test and the Wilcoxon test with the same data: $p = 0.001115$ and 0.002988 respectively. The non-parametric test is much more appropriate than the t-test when the errors are not normal, and the non-parametric test is about 95% as powerful with normal errors, and can be **more** powerful than the t-test if the distribution is strongly skewed by the presence of outliers. Typically, as here, the t-test will give the lower p value, so the Wilcoxon test is said to be conservative; if a difference is significant under a Wilcoxon test it would have been even more significant under a t-test.

Tests on Paired Samples

Sometimes, two-sample data come from paired observations. In this case, we might expect a correlation between the two measurements, either because they were made on the same individual, or were taken from the same location. You might recall that earlier (Box 6.1) we found that the variance of a difference was the average of

$$(y_A - \mu_A)^2 + (y_B - \mu_B)^2 - 2(y_A - \mu_A)(y_B - \mu_B),$$

which is the variance of sample A, plus the variance of sample B, minus two times the covariance of A and B. When the covariance of A and B is **positive**, this is a great help because it reduces the variance of the difference, which makes it easier to detect significant differences between the means. Pairing is not always effective, because the correlation between y_A and y_B may be weak.

The following data are a composite biodiversity score based on a kick sample of aquatic invertebrates.

```
streams < -read.table("c:\\temp\\streams.txt",header = T)
attach(streams)
names(streams)
```

```
[ 1]  "down"  "up"
```

The elements are paired because the two samples were taken on the same river, one upstream and one downstream from the same sewage outfall. If we ignore the fact that the samples are paired, it appears that the sewage outfall has no impact on the biodiversity score ($p = 0.6856$):

```
t.test(down,up)
```

```
      Welch Two Sample t-test
data: down and up
t = -0.4088, df = 29.755, p-value = 0.6856
alternative hypothesis: true difference in means is not equal to 0
95 percent confidence interval:
  -5.248256 3.498256
sample estimates:
mean of x mean of y
   12.500    13.375
```

However, if we allow that the samples are paired (simply by specifying the option paired = T), the picture is completely different.

t.test(down,up,paired = T)

```
      Paired t-test
data: down and up
t = -3.0502, df = 15, p-value = 0.0081
alternative hypothesis: true difference in means is not equal to 0
95 percent confidence interval:
  -1.4864388 -0.2635612
sample estimates:
mean of the differences
               -0.875
```

Now, the difference between the means is highly significant ($p = 0.0081$). The moral is clear. If you can do a paired *t*-test, then you should always do the paired test. It can never do any harm, and sometimes (as here) it can do a huge amount of good. In general, if you have information on **blocking** or **spatial correlation** (in this case, the fact that the two samples came from the same river), then you should always use it in the analysis.

Here is the same paired test carried out as a one-sample *t*-test on the differences between the pairs:

d <- up-down
t.test(d)

```
      One Sample t-test
data: d
t = 3.0502, df = 15, p-value = 0.0081
alternative hypothesis: true mean is not equal to 0
95 percent confidence interval:
  0.2635612 1.4864388
sample estimates:
mean of x
   0.875
```

As you see, the result is identical to the two-sample t-test with paired $= T$ ($p = 0.0081$). The upstream values of the biodiversity score were greater by 0.875 on average, and this difference is highly significant. Working with the differences has halved the number of degrees of freedom (from 30 to 15), but it has more than compensated for this by reducing the error variance, because there is such a strong positive correlation between y_A and y_B.

The Sign Test

This is one of the simplest of all statistical tests. Suppose that you cannot **measure** a difference, but you can **see** it (e.g. in judging a diving contest). For example, nine springboard divers were scored as better or worse, having trained under a new regime and under the conventional regime (the regimes were allocated in a randomized sequence to each athlete: new then conventional, or conventional then new). Divers were judged twice – one diver was worse on the new regime, and eight were better. What is the evidence that the new regime produces significantly better scores in competition? The answer comes form a two-tailed binomial test. How likely is a response of 1/9 (or 8/9 or more extreme than this, i.e. 0/9 or 9/9) if the populations are actually the same (i.e. $p = 0.5$) ? We use a binomial test for this, specifying the number of 'failures' (1) and the total sample size (9):

binom.test(1,9)

This produces the output

```
        Exact binomial test

data: 1 out of 9
number of successes = 1, n = 9, p-value = 0.0391
alternative hypothesis: true p is not equal to 0.5
```

from which we would conclude that the new training regime is significantly better than the traditional method, because $p < 0.05$.

It is easy to write a function to carry out a sign test to compare two samples, x and y

```
sign.test <- function(x, y)
{
if(length(x) != length(y)) stop("The two variables must be the same length")
d <- x - y
binom.test(sum(d > 0), length(d))
}
```

The function starts by checking that the two vectors are the same length, then works out the vector of the differences, d. The binomial test is then applied to the number of positive differences (sum(d > 0)) and the total number of numbers (length(d)). If there was no difference between the samples then, on average, the sum would be about half of length(d). Here is the sign test used to compare the ozone levels in gardens A and B:

sign.test(gardenA,gardenB)

```
     Exact binomial test

data: sum(d > 0) and length(d)
number of successes = 0, number of trials = 10, p-value = 0.001953
alternative hypothesis: true probability of success is not equal to 0.5
95 percent confidence interval:
0.0000000 0.3084971
sample estimates:
probability of success
                    0
```

Note that the p value (0.002) from the sign test is larger than in the equivalent t-test ($p = 0.0011$) that we carried out earlier. This will generally be the case: other things being equal, the parametric test will be more powerful than the non-parametric equivalent.

Binomial Tests to Compare Two Proportions

Suppose that only four females were promoted compared with 196 men. Is this an example of blatant sexism, as it might appear at first glance? Before we can judge, of course, we need to know the number of male and female candidates. It turns out that 196 men were promoted out of 3270 candidates, compared with four promotions out of only 40 candidates for the women. Now, if anything, it looks like the females did better than males in the promotion round (10% success for women versus 6% success for men).

The question then arises as to whether the apparent positive discrimination in favour of women is statistically significant, or whether this sort of difference could arise through chance alone. This is easy in R using the built-in binomial proportions test prop.test in which we specify two vectors, the first containing the number of successes for females and males c(4,196) and second containing the total number of female and male candidates c(40,3270)

prop.test(c(4,196),c(40,3270))

```
2-sample test for equality of proportions with continuity correction

data: c(4, 196) out of c(40, 3270)
X-squared = 0.5229, df = 1, p-value = 0.4696
alternative hypothesis: two.sided
95 percent confidence interval:
 -0.06591631 0.14603864
sample estimates:
    prop 1      prop 2
0.10000000 0.05993884
```

There is no evidence in favour of positive discrimination ($p = 0.4696$). A result like this will occur more than 45% of the time by chance alone. Just think what would have happened if one of the successful female candidates had not applied. Then the same promotion system would have produced a female success rate of 3/39 instead of 4/40 (7.7% instead of 10%). In small samples, small changes have big effects.

Chi-square Contingency Tables

A great deal of statistical information comes in the form of **counts** (whole numbers or integers): the number of animals that died, the number of branches on a tree, the number of days of frost, the number of companies that failed, the number of patients that died. With count data, the number 0 is often the value of a response variable (consider, for example, what a 0 would mean in the context of the examples just listed).

The dictionary definition of contingency is 'a thing dependent on an uncertain event' (OED 2004). In statistics, however, the contingencies are **all the events that could possibly happen**. A contingency table shows the counts of how many times each of the contingencies actually happened in a particular sample. Consider the following example that has to do with the relationship between hair colour and eye colour in white people. For simplicity, we just chose two contingencies for hair colour: 'fair' and 'dark'. Likewise we just chose two contingencies for eye colour: 'blue' and 'brown'. These two categorical variables, eye colour and hair colour, each has two levels ('blue' and 'brown', and 'fair' and 'dark' respectively). Between them, they define four possible outcomes (the contingencies): fair hair and blue eyes, fair hair and brown eyes, dark hair and blue eyes, and dark hair and brown eyes. We take a sample of people and count how many of them fall into each of these four categories. Then we fill in the two-by-two contingency table:

	Blue eyes	Brown eyes
Fair hair	38	11
Dark hair	14	51

These are our observed frequencies (or counts). The next step is very important. In order to make any progress in the analysis of these data we need a **model** which predicts the expected frequencies. What would be a sensible model in a case like this? There are all sorts of complicated models that you might select, but the simplest model (Occam's razor, or the Principle of Parsimony) is that hair colour and eye colour are **independent**. We may not believe that this is actually true, but the hypothesis has the great virtue of being falsifiable. It is also a very sensible model to choose because it makes it easy to predict the expected frequencies based on the assumption that the model is true. We need to do some simple probability work. What is the probability of getting a random individual from this sample whose hair was fair? A total of 49 people (38 + 11) had fair hair out of a total sample of 114 people. So the probability of fair hair is 49/114 and the probability of dark hair is 65/114. Notice that because we have only two levels of hair colour, these two probabilities add up to one [(49 + 65)/114]. What about eye colour? What is the probability of selecting someone at random from this sample with blue eyes? A total of 52 people had blue eyes (38 + 14) out of the sample of 114, so the probability of blue eyes is 52/114 and the probability of brown eyes is 62/114. As before, these add up to one [(52 + 62)/114]. It helps to add the subtotals to the margins of the contingency table like this:

	Blue eyes	Brown eyes	Row totals
Fair hair	38	11	49
Dark hair	14	51	65
Column totals	52	62	114

Now comes the important bit. We want to know the expected frequency of people with fair hair *and* blue eyes, to compare with our observed frequency of 38. Our model says that the two are independent. This is essential information, because it allows us to calculate the expected probability of fair hair and blue eyes. **If, and only if, the two traits are independent, then the probability of having fair hair and blue eyes is the product of the two probabilities**. So, following our earlier calculations, the probability of fair hair and blue eyes is $49/114 \times 52/114$. We can do exactly equivalent things for the other three cells of the contingency table:

	Blue eyes	Brown eyes	Row totals
Fair hair	$\frac{49}{114} \times \frac{52}{114}$	$\frac{49}{114} \times \frac{62}{114}$	49
Dark hair	$\frac{65}{114} \times \frac{52}{114}$	$\frac{65}{114} \times \frac{62}{114}$	65
Column totals	52	62	114

Now we need to know how to calculate the expected frequency. It couldn't be simpler. It is just the probability multiplied by the total sample ($n = 114$). So the expected frequency of blue eyes and fair hair is $\frac{49}{114} \times \frac{52}{114} \times 114 = 22.35$ which is much less than our observed frequency of 38. It is beginning to look as if our hypothesis of independence of hair and eye colour is false.

You might have noticed something useful in the last calculation: two of the sample sizes cancel out. Therefore, the expected frequency in each cell is just the row total (R) times the column total (C) divided by the grand total (G) like this:

$$E = \frac{R \times C}{G}.$$

We can now work out the four expected frequencies.

	Blue eyes	Brown eyes	Row totals
Fair hair	22.35	26.65	49
Dark hair	29.65	35.35	65
Column totals	52	62	114

Notice that the row and column totals (the so-called 'marginal totals') are retained under the model. It is clear that the observed frequencies and the expected frequencies are different, but in sampling, everything always varies, so this is no surprise. The important question is whether or not the expected frequencies are **significantly** different from the observed frequencies.

We assess the significance of the differences between the observed and expected frequencies using a Chi-square test. We calculate a test statistic χ^2 (Pearson's chi square) as follows:

$$\chi^2 = \sum \frac{(O - E)^2}{E},$$

where O is the observed frequency and E is the expected frequency. Capital Greek sigma $\sum$ just means 'add up all the values of'. It makes the calculations easier if we write the observed and expected frequencies in parallel columns, so that we can work out the corrected squared differences more easily.

	O	E	$(O\text{-}E)^2$	$\frac{(O-E)^2}{E}$
Fair hair and blue eyes	38	22.35	244.92	10.96
Fair hair and brown eyes	11	26.65	244.92	9.19
Dark hair and blue eyes	14	29.65	244.92	8.26
Dark hair and brown eyes	51	35.35	244.92	6.93

All we need to do now is to add up the four components of chi square to get $\chi^2 = 35.33$. The question now arises: is this a big value of chi square or not? This is important, because if it **is** a bigger value of chi square than we would expect by chance, then we should reject the null hypothesis. If, on the other hand, it is within the range of values that we would expect by chance alone, then we should accept the null hypothesis.

We always proceed in the same way at this stage. We have a calculated value of the test statistic: $\chi^2 = 35.33$. We compare this value of the test statistic with the relevant critical value. To work out the critical value of chi square we need two things:

- the number of degrees of freedom, and

- the degree of certainty with which to work.

In general, a contingency table has a number of rows (r) and a number of columns (c), and the degrees of freedom are given by

$$\text{d.f.} = (r - 1) \times (c - 1).$$

So we have $(2 - 1) \times (2 - 1) = 1$ degree of freedom for a 2×2 contingency table. You can see why there is only one degree of freedom by working through our example. Take the 'fair hair, brown eyes' box (the top right in the table) and ask 'how many values could this possibly take'? The first thing to note is that the count could not be more than 49, otherwise the row total would be wrong but, in principle, the number in this box is free to be any value between 0 and 49. We have one degree of freedom for this box. But when we have fixed this box to be 11

	Blue eyes	Brown eyes	Row totals
Fair hair		11	49
Dark hair			65
Column totals	52	62	114

you will see that we have no freedom at all for any of the other three boxes. The top left box has to be $49 - 11 = 38$ because the row total is fixed at 49. Once the top left box is defined as 38 then the bottom left box has to be $52 - 38 = 14$ because the column total is fixed (the total number of people with blue eyes was 52). This means that the bottom right box has to be $65 - 14 = 51$. Thus, because the marginal totals are constrained, a 2×2 contingency table has just one degree of freedom.

The next thing we need to do is say how certain we want to be about the falseness of the null hypothesis. The more certain we want to be, the larger the value of chi square we would need to reject the null hypothesis. It is conventional to work at the 95% level. That is our certainty level, so our uncertainty level is $100 - 95 = 5\%$. Expressed as a fraction, this is called alpha ($\alpha = 0.05$). Technically, alpha is the probability of **rejecting** the null hypothesis when it is **true**. This is called a Type I error. A Type II error is **accepting** the null hypothesis when it is **false**.

Critical values in R are obtained by use of *quantiles* (q) of the appropriate statistical distribution. For the chi-squared distribution, this function is called qchisq. The function has two arguments: the certainty level ($p = 0.95$), and the degrees of freedom (d.f. $= 1$):

```
qchisq(0.95,1)
```

```
[1] 3.841459
```

The critical value of chi squared is 3.841. The logic goes like this: since the calculated value of the test statistic is **greater** than the critical value we **reject** the null hypothesis. You should memorize this sentence and put the emphasis on 'greater' and 'reject'.

What have we learned so far? We have rejected the null hypothesis that eye colour and hair colour are independent. However, that's not the end of the story, because we have not established the **way** in which they are related (e.g. is the correlation between them positive or negative?). To do this we need to look carefully at the data and compare the observed and expected frequencies. If fair hair and blue eyes were positively correlated, would the observed frequency be greater or less than the expected frequency? A moment's thought should convince you that the observed frequency will be greater than the expected frequency when the traits are positively correlated (and less when they are negatively correlated). In our case we expected only 22.35 but we observed 38 people (nearly twice as many) to have both fair hair and blue eyes. So it is clear that fair hair and blue eyes are **positively** associated.

In R the procedure is very straightforward. We start by defining the counts as a 2 × 2 matrix like this:

```
count < -matrix(c(38,14,11,51),nrow = 2)
count
```

```
      [ ,1]     [ ,2]
[ 1,]   38        11
[ 2,]   14        51
```

Notice that you enter the data **column-wise** (not row-wise) into the matrix. Then the test uses the chisq.test function, with the matrix of counts as its only argument.

```
chisq.test(count)
```

```
        Pearson' s Chi-squared test with Yates' continuity correction

data: count
X-squared = 33.112, df = 1, p-value = 8.7e-09
```

The calculated value of chi square is slightly different from ours, because Yates' correction has been applied as the default (see Sokal and Rohlf 1995: p. 736). If you switch the correction off (correct = F), you get the value we calculated by hand:

```
chisq.test(count,correct = F)
```

```
        Pearson' s Chi-squared test

data: count
X-squared = 35.3338, df = 1, p-value = 2.778e-09
```

It makes no difference at all to the interpretation that there is a highly significant positive association between fair hair and blue eyes for this group of people.

Fisher's Exact Test

This test is used for the analysis of contingency tables in which **one or more of the expected frequencies is less than 5**. The individual counts are a, b, c and d:

2 × 2 Table	Column 1	Column 2	Row totals
Row 1	a	b	$a + b$
Row 2	c	d	$c + d$
Column totals	$a + c$	$b + d$	n

The probability of any one particular outcome is given by

$$p = \frac{(a+b)!(c+d)!(a+c)!(b+d)!}{a!b!c!d!n!},$$

where n is the grand total, and ! means 'factorial' (the product of all the numbers from n down to 1; zero! is defined as being 1).

Our data concern the distribution of eight ants' nests over ten trees of each of two species (A and B). There are two categorical explanatory variables (ants and trees), and four contingencies, ants (present or absent) and trees (A or B). The response variable (shaded cells) is the vector of four counts c(6,4,2,8).

	Tree A	Tree B	Row totals
With ants	6	2	8
Without ants	4	8	12
Column totals	10	10	20

R does not have a function to calculate factorials, but we can easily write one based on the maximum value of the cumulative product of the numbers from 1 to x:

```
factorial < -function(x) max(cumprod(1:x))
```

Now we can calculate the probability for this particular outcome:

```
factorial(8)*factorial(12)*factorial(10)*factorial(10)/(factorial(6)
            *factorial(2)*factorial(4)*factorial(8)*factorial(20))
```

```
[ 1] 0.07501786
```

This is only part of the story. We need to compute the probability of outcomes that are **more extreme** than this. There are two of them. Suppose only one ant colony was found on Tree B. Then the table values would be 7, 1, 3, 9 but the row and column totals would be exactly the same (the marginal totals are constrained). The numerator always stays the same, so this case has probability

factorial(8)*factorial(12)*factorial(10)*factorial(10)/
                (factorial(7)*factorial(3)*factorial(1)*factorial(9)*factorial(20))

[1] 0.009526078

There is an even more extreme case if no ant colonies at all were found on Tree B. Now the table elements become 8, 0, 2, 10 with probability

factorial(8)*factorial(12)*factorial(10)*factorial(10)/
                (factorial(8)*factorial(2)*factorial(0)*factorial(10)*factorial(20))

[1] 0.0003572279

and we need to add these three probabilities together

0.07501786 + 0.009526078 + 0.000352279

[1] 0.08489622

However, there was no *a priori* reason for expecting the result to be in this direction. It might have been Tree A that had relatively few ant colonies. We need to allow for extreme counts in the opposite direction by doubling this probability (all Fisher's Exact Tests are two-tailed).

2*(0.07501786 + 0.009526078 + 0.000352279)

[1] 0.1697924

This shows that there is no evidence of a correlation between tree and ant colonies. The observed pattern, or a more extreme one, could have arisen by chance alone with probability $p = 0.17$.

There is a built-in function called fisher.test, which saves us all this tedious computation. It takes as its argument a 2×2 matrix containing the counts of the four contingencies. We make the matrix like this (compare with the alternative method of making a matrix, above):

```
x <-as.matrix(c(6,4,2,8))
dim(x) <-c(2,2)
x
```

```
       [,1]  [,2]
[1,]     6     2
[2,]     4     8
```

and run the test like this

fisher.test(x)

```
        Fisher's Exact Test for Count Data

data: x
p-value = 0.1698
alternative hypothesis: true odds ratio is not equal to 1
95 percent confidence interval:
  0.6026805 79.8309210
sample estimates:
odds ratio
  5.430473
```

The fisher.test can be used with matrices much bigger than 2×2. Alternatively, the function may be provided with two vectors containing factor levels, instead of a two-dimensional matrix of counts, as here; this saves you the trouble of counting up how many combinations of each factor level there are:

table < -read.table("c:\\temp\\fisher.txt",header = T)
table

```
        tree      nests
1        A        ants
2        B        ants
3        A        none
4        A        ants
5        B        none
6        A        none
7        A        ants
8        B        ants
9        B        none
10       A        none
11       A        none
12       B        none
13       B        none
14       A        ants
15       A        ants
16       B        none
17       A        ants
18       B        none
19       B        none
20       B        none
```

attach(table)
fisher.test(tree,nests)

```
    Fisher' s Exact Test for Count Data
data: tree and nests
p-value = 0.1698
alternative hypothesis: true odds ratio is not equal to 1
95 percent confidence interval:
  0.6026805 79.8309210
sample estimates:
odds ratio
  5.430473
```

Correlation and Covariance

With two continuous variables, x and y, the question naturally arises as to whether their values are correlated with each other. Correlation is defined in terms of the variance of x, the variance of y, and the covariance of x and y (the way the two vary together, or the way they co-vary) on the assumption that both variables are normally distributed. We have symbols already for the two variances; s_x^2 and s_y^2. Now we call the covariance of x and y $\mathrm{cov}(x,y)$, after which the correlation coefficient r is defined as

$$r = \frac{\mathrm{cov}(x, y)}{\sqrt{s_x^2.s_y^2}}.$$

We know how to calculate variances, so it remains only to work out the value of the covariance of x and y. Covariance is defined as **the expectation of the vector product** $x * y$ which sounds difficult, but isn't (Box 6.2). The covariance of x and y is 'the expectation of the product minus the product of the two expectations'. Note that when x and y are independent (i.e. they are not correlated) then the covariance between x and y is 0, so $\mathbf{E}[xy] = \mathbf{E}[x].\mathbf{E}[y]$ (i.e. the product of their mean values).

Box 6.2 Correlation and covariance

The correlation coefficient is defined in terms of the covariance of x and y, and the geometric mean of the variances of x and y:

$$\rho(x, y) = \frac{\mathrm{cov}(x, y)}{\sqrt{\mathrm{var}(x) \times \mathrm{var}(y)}}.$$

We know how to compute $\mathrm{var}(x)$ and $\mathrm{var}(y)$, so we need only to find $\mathrm{cov}(x,y)$. The covariance of x and y is defined as the **expectation** of the vector product $(x - \bar{x})(y - \bar{y})$

$$\mathrm{cov}(x, y) = \mathbf{E}[(x - \bar{x})(y - \bar{y})].$$

We start by multiplying through the brackets:

$$(x - \bar{x})(y - \bar{y}) = xy - \bar{x}y - x\bar{y} + \overline{xy}.$$

Now applying expectations, and remembering that the expectation of x is $\bar{x}$ and the expectation of y is $\bar{y}$ we get

$$\mathrm{cov}(x, y) = \mathbf{E}(xy) - \bar{x}\mathbf{E}(y) - \mathbf{E}(x)\bar{y} + \bar{x}\bar{y} = \mathbf{E}(xy) - \bar{x}\bar{y} - \bar{x}\bar{y} + \bar{x}\bar{y}.$$

Then $-\bar{x}\bar{y} + \bar{x}\bar{y}$ cancels out, leaving $-\bar{x}\bar{y}$ which is $-\mathbf{E}(x)\mathbf{E}(y)$ so

$$\mathrm{cov}(x, y) = \mathbf{E}(xy) - \mathbf{E}(x)\mathbf{E}(y).$$

Notice that when x and y are uncorrelated, $\mathbf{E}(xy) = \mathbf{E}(x)\mathbf{E}(y)$ so the covariance is 0 in this case. The corrected sum of products $SSXY$ (see p. 133) is given by

$$SSXY = \sum xy - \frac{\sum x \sum y}{n},$$

so covariance is computed as:

$$\mathrm{cov}(x, y) = SSXY \sqrt{\frac{1}{(n-1)^2}}$$

$SSXY$ also provides a shortcut formula for the correlation coefficient

$$r = \frac{SSXY}{\sqrt{SSX.SSY}}$$

because the degrees of freedom $(n - 1)$ cancel out. The sign of r takes the sign of $SSXY$: positive for positive correlations and negative for negative correlations.

Let's do a numerical example.

```
data < -read.table("c:\\temp\\twosample.txt",header = T)
attach(data)
plot(x,y)
```

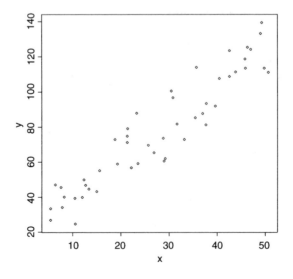

First we need the variance of x and the variance of y:

var(x)

```
[ 1]  199.9837
```

var(y)

```
[ 1]  977.0153
```

The covariance of x and y, cov(x,y), is given by the var function when we supply it with two vectors like this:

var(x,y)

```
[ 1]  414.9603
```

Thus, the correlation coefficient should be $414.96/\sqrt{199.98 \times 977.02}$

var(x,y)/sqrt(var(x)*var(y))

```
[ 1]  0.9387684
```

Let's see if this checks out:

cor(x,y)

```
[ 1]  0.9387684
```

Yes it does! So now you know the definition of the correlation coefficient: it is the covariance divided by the geometric mean of the two variances.

Data Dredging

The R function cor returns the correlation matrix of a data matrix, or a single value showing the correlation between one vector and another.

pollute < -read.table("c:\\temp\\pollute.txt",header = T)
attach(pollute)
names(pollute)

```
[ 1]   "Pollution"   "Temp"      "Industry"   "Population"   "Wind"
[ 6]   "Rain"        "Wet.days"
```

cor(pollute)

	Pollution	Temp	Industry	Population	Wind
Pollution	1.00000000	−0.43360020	0.64516550	0.49377958	0.09509921
Temp	−0.43360020	1.00000000	−0.18788200	−0.06267813	−0.35112340
Industry	0.64516550	−0.18788200	1.00000000	0.95545769	0.23650590

Population	0.49377958	−0.06267813	0.95545769	1.00000000	0.21177156
Wind	0.09509921	−0.35112340	0.23650590	0.21177156	1.00000000
Rain	0.05428389	0.38628047	−0.03121727	−0.02606884	−0.01246601
Wet.days	0.36956363	−0.43024212	0.13073780	0.04208319	0.16694974

	Rain	Wet.days
Pollution	0.05428389	0.36956363
Temp	0.38628047	−0.43024212
Industry	−0.03121727	0.13073780
Population	−0.02606884	0.04208319
Wind	−0.01246601	0.16694974
Rain	1.00000000	0.49605834
Wet.days	0.49605834	1.00000000

The phrase 'data dredging' is used disparagingly to describe the act of trawling through a table like this, desperately looking for big values which might suggest relationships that you can publish. This behaviour is not to be encouraged. The correct approach is model simplification (see p. 195). Note that the correlations are identical in opposite halves of the matrix (in contrast to regression, where regression of y on x would be different from a regression of x on y). The correlation between two vectors produces a single value:

cor(Pollution,Wet.days)

[1] 0.3695636

Correlations with single explanatory variables can be highly misleading if (as is typical) there is substantial correlation amongst the explanatory variables (see Chapter 11).

Partial Correlation

With more than two variables, you often want to know the correlation between x and y when a third variable, say z, is held constant. The **partial correlation coefficient** measures this. It enables correlation due to a shared common cause to be distinguished from direct correlation:

$$r_{xy.z} = \frac{r_{xy} - r_{xz} \cdot r_{yz}}{\sqrt{(1 - r_{xz}^2)(1 - r_{yz}^2)}}.$$

Suppose we had four variables and we wanted to look at the correlation between x and y holding the other two, z and w, constant

$$r_{xy.zw} = \frac{r_{xy.z} - r_{xw.z} \cdot r_{yw.z}}{\sqrt{(1 - r_{xw.z}^2)(1 - r_{yw.z}^2)}}.$$

You will need partial correlation coefficients (p.c.c.) if you want to do **path analysis**. In this book, we prefer to use tree models and various kinds of model simplification following multiple regression. Nevertheless, if you need them, you can use the built-in function lm to get the values of partial correlation coefficients as follows. The sum of squares attributable to a given variable can be determined by deleting it from a model containing all the other variables, using update with anova. Divide this sum of squares by *SSY* and you get what you might call a partial r^2. Take the square root of this to get a partial correlation coefficient.

Correlation and the Variance of Differences Between Variables

Samples often exhibit positive correlations that result from the pairing, as in the upstream and downstream invertebrate biodiversity data that we investigated earlier. There is an important general question about the effect of correlation on the variance of differences between variables. In the extreme, when two variables are so perfectly correlated that they are identical, then the difference between one variable and the other is zero. So it is clear that the variance of a difference will decline as the strength of positive correlation increases.

The following data show the depth of the water table (m below the surface) in winter and summer at nine locations:

```
paired < -read.table("c:\\temp\\paired.txt",header = T)
attach(paired)
names(paired)
```

```
[ 1] "Location" "Summer" "Winter"
```

We begin by asking whether there is a correlation between summer and winter water table depths across locations:

```
cor(Summer, Winter)
```

```
[ 1] 0.8820102
```

There is a strong positive correlation. Not surprisingly, places where the water table is high in summer tend to have a high water table in winter as well. If you want to determine the significance of a correlation (i.e. the *p* value associated with the calculated value of *r*) then use cor.test rather than cor. This test has non-parametric options for Kendall's tau or Spearman's rank depending on the method you specify (method = "k" or method = "s"), but the default method is Pearson's product–moment correlation (method = "p"):

```
cor.test(Summer, Winter)
```

```
        Pearson' s product-moment correlation
data: Summer and Winter
t = 4.9521, df = 7, p-value = 0.001652
alternative hypothesis: true correlation is not equal to 0
95 percent confidence interval:
  0.5259984 0.9750087
sample estimates:
       cor
0.8820102
```

The correlation is highly significant ($p = 0.00165$). Now, let's investigate the relationship between the correlation coefficient and the three variances: the summer variance, the winter variance, and **the variance of the differences** (summer–winter)

```
varS = var(Summer)
varW = var(Winter)
varD = var(Summer-Winter)
```

The correlation coefficient ρ is related to these three variances by:

$$\rho = \frac{\sigma_y^2 + \sigma_z^2 - \sigma_{y-z}^2}{2\sigma_y\sigma_z}.$$

So, using the values we have just calculated, we get the correlation coefficient to be

```
(varS + varW-varD)/(2*sqrt(varS)*sqrt(varW))
```

```
[1] 0.8820102
```

which checks out. We can also see whether the variance of the difference is equal to the sum of the component variances (see p. 76):

```
varD
```

```
[1] 0.01015
```

```
varS + varW
```

```
[1] 0.07821389
```

No, it is not. They would be equal only if the two samples were independent. In fact, we know that the two variables are positively correlated, so the variance of the difference should be **less** than the sum of the variances by an amount equal to $2 \times r \times s_1 \times s_2$

```
varS + varW -2 * 0.8820102 * sqrt(varS) * sqrt(varW)
```

```
[1] 0.01015
```

which is a better result.

Scale-dependent Correlations

Another major difficulty with correlations is that scatterplots can give a highly misleading impression of what is going on. The moral of this exercise is very important: **things are not always as they seem**. The data show the number of species of mammals in forests of differing productivity:

```
par(mfrow = c(1,1))
rm(x,y)
productivity < -read.table("c:\\temp\\productivity.txt",header = T)
attach(productivity)
names(productivity)
```

```
[1] "x" "y" "f"
```

```
plot(x,y,ylab = "Mammal species",xlab = "Productivity")
```

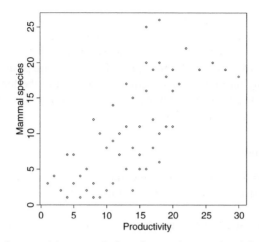

There is a very clear positive correlation: increasing productivity is associated with increasing species richness. The correlation is highly significant:

cor.test(x,y,method = "spearman")

```
        Spearman's rank correlation rho
data: x and y
S = 6515, p-value = < 2.2e-16
alternative hypothesis: true rho is not equal to 0
sample estimates:
    rho
0.7516389
```

However, what if we look at the relationship for each region separately, using **xyplot** from the library of lattice plots (see the web site)?

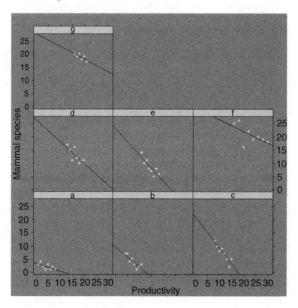

I've added the regression lines for emphasis, but the pattern is obvious. In every single case, increasing productivity is associated with **reduced** mammal species richness within each region (labelled a–g). The lesson is clear: you need to be extremely careful when looking at **correlations across different scales**. Things that are positively correlated over short time scales may turn out to be negatively correlated in the long term. Things that appear to be positively correlated at large spatial scales may turn out (as in this example) to be negatively correlated at small scales.

Kolmogorov–Smirnov Test

People know this test for its wonderful name, rather than for what it actually does. It is an extremely simple test for asking one of two different questions.

- Are two sample distributions the same, or are they significantly different from one another?

- Does a particular sample distribution arise from a particular hypothesized distribution?

The two-sample problem is the one most often used. The apparently simple question is actually very broad. It is obvious that two distributions could be different because their means were different; but two distributions with exactly the same mean could be significantly different if they differed in variance, or in skew or kurtosis (see p. 69). The Kolmogorov–Smirnov test works on **cumulative distribution functions (c.d.f.)**. These give the probability that a randomly selected value of $X \leq x$

$$F(x) = P[X \leq x]$$

This sounds somewhat abstract. Suppose we had insect wing sizes for two geographically separated populations and we wanted to test whether the distribution of wing lengths was the same in the two places.

```
wings < -read.table("c:\\temp\\wings.txt",header = T)
attach(wings)
names(wings)
```

```
[ 1] "size" "location"
```

We need to find out how many specimens there are from each location:

```
table(location)
location
 A   B
50  70
```

So the samples are of unequal size (50 insects from location A, 70 from B). It will be useful, therefore, to create two separate variables containing the wing lengths from sites A and B:

```
A < -size[location = = "A"]
B < -size[location = = "B"]
```

We could begin by comparing mean wing length in the two locations with a *t*-test:

t.test(A,B)

```
        Welch Two Sample t-test

data: A and B
t = -1.6073, df = 117.996, p-value = 0.1107
alternative hypothesis: true difference in means is not equal to 0
95 percent confidence interval:
  -2.494476 0.259348
sample estimates:
mean of x mean of y
  24.11748 25.23504
```

This shows that mean wing length is not significantly different in the two locations ($p = 0.11$); but what about other attributes of the distribution? This is where Kolmogorov–Smirnov is really useful:

ks.test(A,B)

```
        Two-sample Kolmogorov-Smirnov test

data: A and B
D = 0.2629, p-value = 0.02911
alternative hypothesis: two.sided
```

The two distributions are, indeed, significantly different from one another ($p < 0.05$); but if not in their means, then in what respect do they differ? Perhaps they have different variances?

var.test(A,B)

```
        F test to compare two variances

data: A and B
F = 0.5014, num df = 49, denom df = 69, p-value = 0.01192
alternative hypothesis: true ratio of variances is not equal to 1
95 percent confidence interval:
  0.3006728 0.8559914
sample estimates:
ratio of variances
        0.5014108
```

Indeed they do: the variance of wing length from location B is double that from location A ($p < 0.02$).

We can finish by drawing the two histograms side by side to get a visual impression of the difference in the shape of the two distributions; the open bars show the data from location B, solid bars show location A (see the web site).

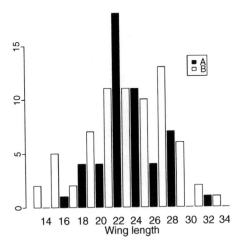

The spread of wing lengths is much greater at location B despite the fact that the mean wing length is similar in the two places. Also, the distribution is skew to the left in location B, with the result that modal wing length is greater in location B (26 mm compared with 22 mm).

7

Statistical Modelling

Fitting models to data is the central function of R. The process is essentially one of exploration; there are no fixed rules and no absolutes. The object is to determine a minimal adequate model from the large set of potential models that might be used to describe the given set of data. In this book we discuss five types of model:

- the null model,
- the minimal adequate model,
- the current model,
- the maximal model, and
- the saturated model.

The step-wise progression from the saturated model (or the maximal model, whichever is appropriate) through a series of simplifications to the minimal adequate model is made on the basis of **deletion tests** – F-tests or chi-squared tests that assess the significance of the increase in deviance that results when a given term is removed from the current model.

Models are representations of reality that should be both accurate and convenient. However, it is impossible to maximize a model's realism, generality and holism simultaneously, and the principle of parsimony (or Occam's razor; see p. 7) is a vital tool in helping to choose one model over another. Thus, we would only include an explanatory variable in a model if it significantly improved the fit of a model. Just because we went to the trouble of measuring something, does not mean we have to have it in our model. Parsimony says that, other things being equal, we prefer:

- a model with $n - 1$ parameters to a model with n parameters,
- a model with $k - 1$ explanatory variables to a model with k explanatory variables,
- a linear model to a model which is curved,
- a model without a hump to a model with a hump, and
- a model without interactions to a model containing interactions between factors.

Statistics: An Introduction using R   M. J. Crawley
© 2005 John Wiley & Sons, Ltd   ISBNs: 0-470-02298-1 (PBK); 0-470-02297-3 (PPC)

Other considerations include a preference for models containing explanatory variables that are easy to measure over variables that are difficult or expensive to measure. Also, we prefer models that are based on a sound mechanistic understanding of the process over purely empirical functions.

Parsimony requires that the model should be as simple as possible. This means that the model should not contain any redundant parameters or factor levels. We achieve this by fitting a maximal model then simplifying it by following one or more of these steps:

- remove non-significant interaction terms,

- remove non-significant quadratic or other non-linear terms,

- remove non-significant explanatory variables,

- group together factor levels that do not differ from one another, and

- in Ancova, set non-significant slopes of continuous explanatory variables to zero.

All the above are subject, of course, to the caveats that the simplifications make good scientific sense, and do not lead to significant reductions in explanatory power.

Just as there is no perfect model, so there may be no optimal scale of measurement for a model. Suppose, for example, we had a process that had Poisson errors with multiplicative effects amongst the explanatory variables. Then, one must choose between three different scales, each of which optimizes one of three different properties:

- the scale of $\sqrt{y}$ would give constancy of variance;

- the scale of $y^{\frac{2}{3}}$ would give approximately normal errors;

- the scale of $\ln(y)$ would give additivity.

Thus, any measurement scale is always going to be a compromise, and you should choose the scale that gives the best overall performance of the model.

Model	Interpretation
Saturated model	One parameter for every data point Fit: perfect Degrees of freedom: none Explanatory power of the model: none
Maximal model	Contains all (p) factors, interactions and covariates that might be of any interest. Many of the model's terms are likely to be insignificant Degrees of freedom: $n - p - 1$ Explanatory power of the model: it depends
Minimal adequate model	A simplified model with $0 \le p' \le p$ parameters Fit: less than the maximal model, but not significantly so Degrees of freedom: $n - p' - 1$

	Explanatory power of the model: $r^2 = SSR/SSY$
Null model	Just one parameter, the overall mean $\bar{y}$
	Fit: none; $SSE = SSY$
	Degrees of freedom: $n - 1$
	Explanatory power of the model: none

The Steps Involved in Model Simplification

There are no hard and fast rules, but the procedure laid out below works well in practice. With large numbers of explanatory variables, and many interactions and non-linear terms, the process of model simplification can take a very long time. However, this is time well spent because it reduces the risk of overlooking an important aspect of the data. It is important to realize that there is no guaranteed way of finding all the important structures in a complex dataframe.

Step	Procedure	Explanation
1	Fit the maximal model	Fit all the factors, interactions and covariates of interest. Note the residual deviance. If you are using Poisson or binomial errors, check for overdispersion and rescale if necessary.
2	Begin model simplification	Inspect the parameter estimates using **summary**. Remove the least significant terms first, using **update -**, starting with the highest-order interactions.
3	If the deletion causes an insignificant increase in deviance	Leave that term out of the model. Inspect the parameter values again. Remove the least significant term remaining.
4	If the deletion causes a significant increase in deviance	Put the term back in the model using **update +**. These are the statistically significant terms as assessed by deletion from the maximal model.
5	Keep removing terms from the model	Repeat steps 3 or 4 until the model contains nothing but significant terms. This is the minimal adequate model. If none of the parameters is significant, then the minimal adequate model is the null model.

Caveats

Model simplification is an important process but it should not be taken to extremes. For example, the interpretation of deviances and standard errors produced with fixed parameters that have been estimated from the data, should be undertaken with caution. Again, the search for 'nice numbers' should not be pursued uncritically. Sometimes there are good scientific reasons for using a particular number (e.g. a power of 0.66 in an allometric relationship between respiration and body mass). It is much more straightforward, for example, to say that yield increases by 2 kg per hectare for every extra unit of fertilizer, than to say that it increases by 1.947 kg. Similarly, it may be preferable to say that the odds of infection increase ten-fold under a given treatment, than to say that the logits increase by 2.321; without model simplification this is equivalent to saying that there is a 10.186-fold increase in the odds. It would be absurd, however, to fix on an estimate of 6 rather than 6.1 just because 6 is a whole number.

Order of Deletion

Remember that **order matters**. If your explanatory variables are correlated with each other, then the significance you attach to a given explanatory variable will depend upon whether you delete it from a maximal model or add it to the null model. If you always test by model simplification then you won't fall into this trap.

The fact that you have laboured long and hard to include a particular experimental treatment does not justify the retention of that factor in the model if the analysis shows it to have no explanatory power. Anova tables are often published containing a mixture of significant and non-significant effects. This is not a problem in orthogonal designs, because sums of squares can be unequivocally attributed to each factor and interaction term. However, as soon as there are missing values or unequal weights, then it is impossible to tell how the parameter estimates and standard errors of the significant terms would have been altered if the non-significant terms had been deleted. The best practice is:

- say whether your data are orthogonal or not,

- present a minimal adequate model,

- give a list of the non-significant terms that were omitted, and the deviance changes that resulted from their deletion.

The reader can then judge for themselves the relative magnitude of the non-significant factors, and the importance of correlations between the explanatory variables.

The temptation to retain terms in the model that are 'close to significance' should be resisted. The best way to proceed is this. If a result would have been **important** if it had been statistically significant, then it is worth repeating the experiment with higher replication and/or more efficient blocking, in order to demonstrate the importance of the factor in a convincing and statistically acceptable way.

Model Formulae in R

The structure of the model is specified in the model formula like this:

$$\textbf{response variable} \sim \textbf{explanatory variable(s)},$$

where the **tilde** symbol $\sim$ reads 'is modelled as a function of'. So a simple linear regression of y on x would be written like this

y~x

and a one-way Anova where sex is a two-level factor would be written like this

y~sex.

The right-hand side of the model formula shows

- the number of explanatory variables and their identities–their attributes (e.g. continuous or categorical) are usually defined prior to the model fit,
- the interactions between the explanatory variables (if any),
- non-linear terms is the explanatory variables.

On the right of the tilde, one also has the option to specify offsets or error terms in some special cases. As with the response variable, the explanatory variables can appear as transformations, or as powers or polynomials.

It is very important to note that symbols are used differently in model formulae than in arithmetic expressions. In particular:

$+$ indicates inclusion of an explanatory variable in the model (not addition);

$-$ indicates deletion of an explanatory variable from the model (not subtraction);

$*$ indicates inclusion of explanatory variables and interactions (not multiplication);

/ indicates nesting of explanatory variables in the model (not division);

| indicates conditioning (e.g. $y \sim x \mid z$ is read as 'y as a function of x given z').

There are several other symbols that have special meanings in model formulae, in particular

: colon means an interaction, so that A:B means the two-way interaction between A and B, and N:P:K:Mg means the four-way interaction between N, P, K and Mg.

Some terms can be written in an expanded form. Thus:

A*B*C is the same as A + B + C + A:B + A:C + B:C + A:B:C

A/B/C is the same as A + B%in%A + C%in%B%in%A

(A + B + C)^3 is the same as A*B*C

(A + B + C)^2 is the same as A*B*C − A:B:C.

Interactions Between Explanatory Variables

Interactions between two two-level categorical variables A*B mean that two main effect means and one interaction mean are evaluated. On the other hand, if factor A has three levels and factor B has four levels, then seven parameters are estimated for the main effects (three means for A and four means for B). The number of interaction terms is $(a - 1)(b - 1)$ where a and b are the numbers of levels of the factors A and B respectively. So in this case, R would estimate $(3 - 1)(4 - 1) = 6$ parameters for the interaction.

Interactions between two continuous variables are fitted differently. If x and z are two continuous explanatory variables, then $x*z$ means fit $x + z + x : z$ and the interaction term $x : z$ behaves as if a new variable had been computed that was the point-wise product of the two vectors x and z. The same effect could be obtained by calculating the product explicitly

```
product.xz <- x * z
```

then using the model formula y ~ x + z + product.xz. Note that the representation of the interaction by the **product** of the two continuous variables is an assumption, not a fact. The real interaction might be of an altogether different functional form (e.g. x * z^2).

Interactions between a categorical variable and a continuous variable are interpreted as an analysis of covariance; a separate slope and intercept are fitted for each level of the categorical variable. So $y \sim A*x$ would fit three regression equations if the factor A had three levels; this would estimate six parameters from the data, three slopes and three intercepts.

The slash operator is used to denote nesting. Thus, with categorical variables A and B

```
y ~ A/B
```

means fit 'A plus B within A'. This could be written in two other equivalent ways:

```
y ~ A + A:B
y ~ A + B %in% A
```

both of which alternatives emphasize that there is no point in attempting to estimate a main effect for B (it is probably just a factor label like 'tree number 1' that is of no scientific interest; see p. 185).

Some functions for specifying non-linear terms and higher-order interactions are useful. To fit a polynomial regression in x and z, we could write

```
y ~ poly(x,3) + poly(z,2)
```

to fit a cubic polynomial in x and a quadratic polynomial in z.

To fit interactions, but only up to a certain level, the ^ operator is useful. This formula

```
y ~ (A + B + C)^2
```

fits all the main effects and two-way interactions (i.e. it excludes the three-way interaction that A*B*C would have included).

The **I** function (capital letter i) stands for 'as is'. It overrides the interpretation of a model symbol as a formula operator when the intention is to use it as an arithmetic operator. Suppose you wanted to fit $1/x$ as an explanatory variable in a regression, you might try this:

y ~ 1/x

but this actually does something very peculiar. It fits x nested within the intercept! When it appears in a model formula, the slash operator is assumed to imply nesting. To obtain the effect we want, we use **I** to write

y ~ I(1/x).

We also need to use **I** when we want * to represent multiplication and ^ to mean 'to the power' rather than an interaction model expansion: thus to fit x and x^2 in a quadratic regression we would write

y ~ x + I(x^2).

Multiple Error Terms

When there is nesting (e.g. split plots in a designed experiment; see p. 177) or temporal pseudoreplication (see p. 13) you can include an error function as part of the model formula. Suppose you had a three-factor factorial experiment with categorical variables A, B and C. The twist is that each treatment is applied to plots of different sizes. A is applied to replicated whole fields, B is applied at random to half fields and C is applied to smaller split–split plots within each field. This is shown in a model formula like this:

y ~ A*B*C + Error(A/B/C).

Note that the terms within the model formula are separated by asterisks to show that it is a full factorial with all interaction terms included, whereas the terms are separated by slashes in the error statement. There are as many terms in the error statement as there are different sizes of plots – three in this case, although the smallest plot size (C in this example) can be omitted from the list – and the terms are listed left to right from the largest to the smallest plots; see p. 176 for details and examples.

The Intercept as Parameter 1

The simple command

y ~ 1

causes the null model to be fit. This works out the grand mean (the overall average) of all the data and works out the total deviance (or the total sum of squares, *SSY*, in models with normal errors and the identity link). In some cases, this may be the minimal adequate model; it is possible that none of the explanatory variables we have measured contribute anything significant to our understanding of the variation in the response variable. This is normally what you don't want to happen at the end of your three-year research project.

To remove the intercept (parameter 1) from a regression model (i.e. to force the regression line through the origin), you fit '-1' like this:

y~x–1.

You should not do this unless you know exactly what you are doing, and exactly why you are doing it (see p. 135 for details). Removing the intercept from an Anova model where all the variables are categorical has a different effect:

y~gender–1.

This gives the mean for males and the mean for females in the summary table, rather than the overall mean and the difference in mean for males (see Contrasts, p. 209).

Update in Model Simplification

In the update function used during model simplification, the dot '.' is used to specify 'what is there already' on either side of the tilde. So if your original model said

model < -lm(y ~A*B)

then the update function to remove the interaction term A:B could be written like this:

model2 < -update(model, ~ . - A:B)

Note that there is no need to repeat the name of the response variable, and the punctuation 'tilde dot' means take model as it is, and remove from it ('minus') the interaction term A:B.

Examples of R Model Formulae

Model	Model formula	Comments
Null	y~1	1 is the intercept in regression models, but here it is the overall mean y
Regression	y~x	x is a continuous explanatory variable
One-way Anova	y~gender	Gender is a two-level categorical variable
Two-way Anova	y~gender + genotype	Genotype is a four-level categorical variable
Factorial Anova	y~N * P * K	N, P and K are two-level factors to be fit along with all their interactions
Three-way Anova	y~N*P*K-N:P:K	As above, but don't fit the three-way interaction

Model	Model formula	Comments
Analysis of covariance	y ~ x + gender	A common slope for y against x but with two intercepts, one for each gender
Analysis of covariance	y ~ x * gender	Two slopes and two intercepts
Nested Anova	y ~ a/b/c	Factor c nested within factor b within factor a
Split-plot Anova	y ~ a*b*c + Error(a/b/c)	A factorial experiment but with three plots sizes and three different error variances, one for each plot size
Multiple regression	y ~ x + z	Two continuous explanatory variables, flat surface fit
Multiple regression	y ~ x * z	Fit an interaction term as well $(x + z + x : z)$
Multiple regression	y ~ x + I(x^2) + z + I(z^2)	Fit a quadratic term for both x and z
Multiple regression	y <- poly(x,2) + z	Fit a quadratic polynomial for x and linear z
Multiple regression	y ~ (x + z + w)^2	Fit three variables plus all their two-way interactions
Non-parametric model	y ~ s(x) + lo(z)	y is a function of smoothed x and loess z
Transformed response and explanatory variables	log(y) ~ I(1/x) + sqrt(z)	All three variables are transformed in the model

Model Formulae for Regression

The important point to grasp is that model formulae look very like equations but there are important differences. Our simplest useful equation looks like this:

$$y = a + bx.$$

It is a two-parameter model with one parameter for the intercept, a, and another for the slope, b, of the graph of the continuous response variable y against a continuous explanatory variable x. The model formula for the same relationship looks like this:

$$y \sim x$$

The equals sign is replaced by a tilde, and all of the parameters are left out. If we had a multiple regression with two continuous explanatory variables x and z the equation would look like this

$$y = a + bx + cz,$$

but the model formula is this

$$y \sim x + z$$

It is all wonderfully simple – but just a minute. How does R know what parameters we want to estimate from the data? We have only told it the names of the explanatory variables. We have said nothing about how to fit them, or what sort of equation we want to fit to the data. The key to this is to understand **what kind of explanatory variable is being fit** to the data. If the explanatory variable x specified on the right of the tilde is a continuous variable, then R **assumes** that you want to do a regression, and hence that you want to estimate two parameters in a linear regression whose equation is $y = a + bx$.

A common misconception is that linear models involve a straight-line relationship between the response variable and the explanatory variables. This is **not** the case, as you can see from these two linear models.

```
par(mfrow=c(1,2))
x<-seq(0,10,0.1)
plot(x,1+x-x^2/15,type="l")
plot(x,3+0.1*exp(x),type="l")
```

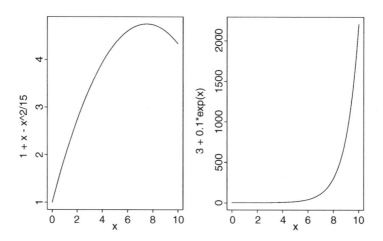

The definition of a linear model is an equation that contains mathematical variables, parameters and random variables that is **linear in the parameters and in the random variables**. What this means is that if a, b and c are parameters then obviously

$$y = a + bx$$

is a linear model, but so is

$$y = a + bx - cx^2$$

because x^2 can be replaced by z which gives a linear relationship

$$y = a + bx + cz$$

and so is

$$y = a + be^x$$

because we can create a new variable $z = \exp(x)$, so that

$$y = a + bz.$$

Some models are non-linear but can be readily linearized by transformation. For example:

$$y = \exp(a + bx)$$

is non-linear, but on taking logs of both sides, it becomes

$$\ln(y) = a + bx.$$

If the equation you want to fit is more complicated than this, then you need to specify the form of the equation, and use non-linear methods (nls or nlme) to fit the model to the data (see p. 149).

GLMs: Generalized Linear Models

We can use glms (pronounced 'glims') when the variance is not constant, and/or when the errors are not normally distributed. Certain kinds of response variables invariably suffer from these two important contraventions of the standard assumptions, and glms are excellent at dealing with them. Specifically, we might consider using glms when the response variable is:

- count data expressed as proportions (e.g. logistic regressions),
- count data that are not proportions (e.g. log linear models of counts),
- binary response variables (e.g. dead or alive), or
- data on time-to-death where the variance increases faster than linearly with the mean (e.g. time data with gamma errors).

The central assumption that we have made up to this point is that variance was constant (a). In count data, however, where the response variable is an integer and there are often lots of zero's in the dataframe, the variance may increase linearly with the mean (b). With proportion data, where we have a count of the number of failures of an event as well as the number of successes, the variance will be a $\cap$-shaped function of the mean (c). Where the response variable follows a gamma distribution (as in data on time-to-death) the

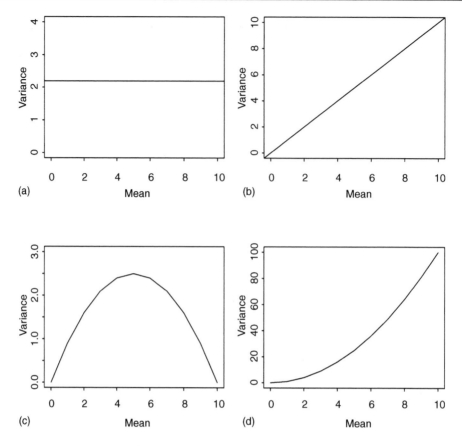

variance increases faster than linearly with the mean (d). Many of the basic statistical methods like regression and Student's t-test assume that variance is constant, but in many applications this assumption is untenable. Hence the great utility of glms.

A generalized linear model has three important properties:

- the **error structure**,

- the **linear predictor**,

- the **link function**.

These are all likely to be unfamiliar concepts. The ideas behind them are straightforward, however, and it is worth learning what each of the concepts involves.

The Error Structure

Up to this point, we have dealt with the statistical analysis of data with normal errors. In practice, however, many kinds of data have non-normal errors. For example:

- errors that are strongly skewed,

- errors that are kurtotic,

- errors that are strictly bounded (as in proportions),

- errors that cannot lead to negative fitted values (as in counts).

In the past, the only tools available to deal with these problems were transformation of the response variable or the adoption of non-parametric methods. A **glm** allows the specification of a variety of different error distributions:

- Poisson errors, useful with count data,

- binomial errors, useful with data on proportions,

- gamma errors, useful with data showing a constant coefficient of variation, and

- exponential errors, useful with data on time-to-death (survival analysis).

The error structure is defined by means of the **family** directive, used as part of the model formula like this:

glm(y~z, family = poisson)

which means that the response variable y has Poisson errors. Or

glm(y~z, family = binomial)

which means that the response is binary, and the model has binomial errors. As with previous models, the explanatory variable z can be continuous (leading to a regression analysis) or categorical (leading to an Anova-like procedure called analysis of deviance, as described below).

The Linear Predictor

The structure of the model relates each observed y-value to a predicted value. The predicted value is obtained **by transformation of the value emerging from the linear predictor**. The linear predictor, η (eta), is a linear sum of the effects of one or more explanatory variables, x_j:

$$\eta_i = \sum_{j=1}^{p} x_{ib}\beta_j,$$

where the x's are the values of the p different explanatory variables, and the β's are the (usually) unknown parameters to be estimated from the data. The right-hand side of the equation is called the **linear structure**.

There are as many terms in the linear predictor as there are parameters, p, to be estimated from the data. Thus with a simple regression, the linear predictor is the sum of two terms whose parameters are the intercept and the slope. With a one-way Anova with four treatments, the linear predictor is the sum of four terms leading to the estimation of the mean for each level of the factor. If there are covariates in the model, they add one term each to the linear predictor (the slope of each relationship). Interaction terms in a factorial Anova add one or more parameters to the linear predictor, depending upon the degrees of freedom of each factor (e.g. there would be three extra parameters for the interaction between a two-level factor and a four-level factor, because $(2 - 1) \times (4 - 1) = 3$).

Fitted Values

To determine the fit of a given model, a **glm** evaluates the linear predictor for each value of the response variable, then compares the predicted value with a **transformed** value of y. The transformation to be employed is specified in the link function, as explained below. The fitted value is computed by applying the reciprocal of the link function, in order to get back to the original scale of measurement of the response variable. Thus, with a log link, the fitted value is the antilog of the linear predictor, and with the reciprocal link, the fitted value is the reciprocal of the linear predictor.

The Link Function

One of the difficult things to grasp about **glm** is the relationship between the values of the response variable (as measured in the data and predicted by the model in fitted values) and the linear predictor. The thing to remember is that the **link function relates the mean value of y to its linear predictor**. In symbols, this means that:

$$\eta = g(\mu)$$

which is simple, but needs thinking about. The linear predictor, η (eta), emerges from the linear model as a sum of the terms for each of the p parameters. **This is not a value of y** (except in the special case of the **identity link** that we have been using (implicitly) up to now). The value of η is obtained by transforming the value of y by the link function, and the predicted value of y is obtained by applying the inverse link function to η.

The most frequently used link functions are shown below. An important criterion in the choice of link function is to ensure that the fitted values stay within reasonable bounds. We would want to ensure, for example, that counts were all greater than or equal to zero (negative count data would be nonsense). Similarly, if the response variable was the proportion of individuals that died, then the fitted values would have to lie between zero and one (fitted values greater than one or less than zero would be meaningless). In the first case, a log link is appropriate because the fitted values are antilogs of the linear predictor, and all antilogs are greater than or equal to zero. In the second case, the logit link is appropriate because the fitted values are calculated as the antilogs of the log-odds, $\log(p/q)$.

By using different link functions, the performance of a variety of models can be compared directly. The total deviance is the same in each case and we can investigate the consequences of altering our assumptions about precisely how a given change in the linear predictor brings about a response in the fitted value of y. The most appropriate link function is the one which produces the minimum residual deviance.

Canonical Link Functions

The canonical link functions are the default options employed when a particular error structure is specified in the **family** directive in the model formula. Omission of a **link** directive means that the following settings are used:

Error	Canonical link
Normal	*identity*
poisson	*log*
binomial	*logit*
Gamma	*reciprocal*

You should try to memorize these canonical links and to understand why each is appropriate to its associated error distribution. Note that only Gamma errors have a capital initial letter.

Choosing between using a link function (e.g. log link) and transforming the response variable (i.e. having $\log(y)$ as the response variable rather than y) takes a certain amount of experience. The decision is usually based on **whether the variance is constant** on the original scale of measurement. If the variance was constant, you would use a link function. If the variance increased with the mean, you would be more likely to log transform the response.

Proportion Data and Binomial Errors

Proportion data have three important properties that affect the way the data should be analysed:

- the data are strictly bounded,
- the variance is non-constant,
- errors are non-normal.

You cannot have a proportion greater than one or less than zero. This has obvious implications for the kinds of functions fitted and for the distributions of residuals around these fitted functions. For example, it makes no sense to have a linear model with a negative slope for proportion data because there would come a point, with high levels of the x variable, that negative proportions would be predicted. Likewise, it makes no sense to have a linear model with a positive slope for proportion data because there would come a point, with high levels of the x variable, that proportions greater than one would be predicted.

With proportion data, if the probability of success is zero, then there will be no successes in repeated trials, all the data will be zeros and hence the variance will be zero.

Likewise, if the probability of success is one, then there will be as many successes
as there are trials, and again the variance will be zero. For proportion data, therefore, the
variance increases with the mean up to a maximum (when the probability of success
is one half) then declines again towards zero as the mean approaches one. The
variance mean relationship is humped, rather than constant as assumed in the classical
tests.

The final assumption is that the errors (the differences between the data and the fitted
values estimated by the model) are normally distributed. This cannot be so in a linear
model because the data are bounded above and below: no matter how big a negative residual at high predicted values, $\hat{y}$, a positive residual cannot be bigger than $1 - \hat{y}$. Similarly,
no matter how big a positive residual might be for low predicted values $\hat{y}$, a negative
residual cannot be greater than $\hat{y}$ (because you cannot have negative proportions). This
means that confidence intervals must be asymmetric whenever $\hat{y}$ takes large values (close
to one) or small values (close to zero).

All these issues (boundedness, non-constant variance, non-Normal errors) are dealt
with by using a generalized linear model with a binomial error structure. It could not be
simpler to deal with this. Instead of using a linear model and writing

lm(y ~ x)

we use a generalized linear model (glm) and specify that the error family is binomial like
this:

glm(y ~ x,family = binomial).

That's all there is to it. In fact, it is even easier than that, because we don't need to write
'family = '

glm(y ~ x,binomial).

Count Data and Poisson Errors

Count data have a number of properties that need to be considered during modelling:

- count data are bounded below (you can't have counts less than zero),

- variance is not constant (variance increases with the mean),

- errors are not normally distributed, and

- the fact that the data are whole numbers (integers) affects the error distribution.

It is very simple to deal with all these issues by using a glm. All we need to write is

glm(y ~ x,poisson)

and the model is fitted with a log link (to ensure that the fitted values are bounded below)
and Poisson errors (to account for the non-normality).

GAMs: Generalized Additive Models

These models are like glms in that they can have different error structures and different link functions to deal with count data or proportion data. What makes them different is that the shape of the relationship between y and a continuous variable x is not specified by some explicit functional form. Instead, non-parametric smoothers are used to describe the relationship. This is especially useful for relationships that exhibit complicated shapes, like hump-shaped curves (see p. 195). The model looks just like a glm, except that the relationships we want to be smoothed are prefixed by **s**: thus, if we had a three-variable multiple regression (three continuous explanatory variables w, x and z) on count data and we wanted to smooth all three explanatory variables, we would write:

model < -gam(y ~ s(w) + s(x) + s(z),poisson)

Model Criticism

There is a temptation to become personally attached to a particular model. Statisticians call this 'falling in love with your model'. It is as well to remember the following truths about models:

- all models are wrong,
- some models are better than others,
- the correct model can never be known with certainty, and
- the simpler the model, the better it is.

There are several ways that we can improve things if it turns out that our present model is inadequate:

- transform the response variable,
- transform one or more of the explanatory variables,
- try fitting different explanatory variables if you have any,
- use a different error structure,
- use non-parametric smoothers instead of parametric functions,
- use different weights for different y values.

All of these are investigated in the coming chapters. In essence, you need a set of tools to establish whether, and how, your model is inadequate. For example, the model might

- predict some of the y values poorly,
- show non-constant variance,
- show non-Normal errors,

- be strongly influenced by a small number of influential data points,

- show some sort of systematic pattern in the residuals, or

- exhibit overdispersion.

Summary of Statistical Models in R

Models are fitted using one of the model-fitting functions as follows.

- **lm**: fits a linear model with normal errors and constant variance; generally this is used for regression analysis using continuous explanatory variables.

- **aov**: fits analysis of variance with normal errors, constant variance and the identity link; generally used for categorical explanatory variables or Ancovas with a mix of categorical and continuous explanatory variables.

- **glm**: fits generalized linear models to data using categorical or continuous explanatory variables, by specifying one of a family of **error structures** (e.g. Poisson for count data or binomial for proportion data) and a particular **link function**.

- **gam**: fits generalized additive models to data with one of a family of error structures (e.g. Poisson for count data or binomial for proportion data) in which the continuous explanatory variables can (optionally) be fitted as arbitrary smoothed functions using non-parametric smoothers rather than specific parametric functions.

- **lme**: fits linear mixed effects models with specified mixtures of fixed effects and random effects and allows for the specification of correlation structure amongst the explanatory variables and autocorrelation of the response variable (e.g. time series effects with repeated measures).

- **nls**: fits a non-linear regression model via least squares, estimating the parameters of a specified non-linear function.

- **nlme**: fits a specified non-linear function in a mixed effects model where the parameters of the non-linear function are assumed to be random effects; allows for the specification of correlation structure amongst the explanatory variables and auto-correlation of the response variable (e.g. time series effects with repeated measures).

- **loess**: fits a local regression model with one or more continuous explanatory variables using non-parametric techniques to produce a smoothed model surface.

- **tree**: fits a regression tree model using binary recursive partitioning whereby the data are successively split along coordinate axes of the explanatory variables so that at any node, the split is chosen that maximally distinguishes the response variable in the left and the right branches. With a categorical response variable, the tree is called a classification tree, and the model used for classification assumes that the response variable follows a multinomial distribution.

For most of these models, a range of **generic functions** can be used to obtain information about the model. The most important and most frequently used are given below.

summary	produces parameter estimates and standard errors from **lm**, and Anova tables from **aov**; this will often determine your choice between **lm** and **aov**. For either **lm** or **aov** you can choose **summary.aov** or **summary · lm** to get the alternative form of output (an Anova table or a table of parameter estimates and standard errors; see p. 212).
plot	produces diagnostic plots for model checking, including residuals against fitted values, influence tests, etc.
anova	a wonderfully useful function for comparing different models and producing Anova tables.
update	used to modify the last model fit; it saves both typing effort and computing time.

Other useful generics include:

coef	the coefficients (estimated parameters) from the model,
fitted	the fitted values, predicted by the model for the values of the explanatory variables included,
resid	the residuals (the differences between measured and predicted values of y),
predict	uses information from the fitted model to produce smooth functions for plotting a line through the scatterplot of your data.

Model Checking

After fitting a model to data we need to investigate how well the model describes the data. In particular, we should look to see if there are any systematic trends in the goodness of fit. For example, does the goodness of fit increase with the observation number, or is it a function of one or more of the explanatory variables? We can work with the raw residuals:

$$\text{residuals} = y - \text{fitted values}.$$

For instance, we should routinely plot the residuals against:

- the fitted values (to look for non-constancy of variance: heteroscedasticity),
- the explanatory variables (to look for evidence of curvature),
- the sequence of data collection (to look for temporal correlation),
- standard normal deviates (to look for non-normality of errors).

Non-constant Variance: Heteroscedasticity

A good model must also account for the variance–mean relationship adequately and produce additive effects on the appropriate scale (as defined by the link function). A plot of standardized residuals against fitted values should look like the sky at night (points scattered at random over the whole plotting region), with no trend in the size or degree of scatter of the residuals. A common problem is that the variance increases with the mean, so that we obtain an expanding, fan-shaped pattern of residuals.

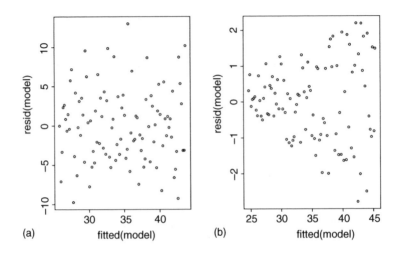

The plot on the left (a) is what we want to see with no trend in the residuals with the fitted values. The plot on the right (b) is a problem. There is a clear pattern of increasing residuals as the fitted values get larger. This is a picture of what **heteroscedasticity** looks like.

Non-Normality of Errors

Errors may be non-Normal for several reasons. They may be skew, with long tails to the left or right. Or they may be kurtotic, with a flatter or more pointy top to their distribution. In any case, the theory is based on the assumption of Normal errors, and if the errors are **not** Normally distributed, then we shall not know how this affects our interpretation of the data or the inferences we make from it.

It takes considerable experience to interpret the Normal error plots. Here we generate a series of data sets where we introduce different but known kinds of non-Normal errors. Then we plot them using a simple home-made function called mcheck (first developed by John Nelder in the original GLIM language; the name stands for model checking). The idea is to see what patterns are generated in Normal plots by the different kinds of non-Normality. In real applications we would use the generic plot(model) rather than mcheck (see below). First, we write the function mcheck. The idea is to produce two plots, side by side – a plot of the residuals against the fitted values on the left, and a plot of the ordered residuals against the quantiles of the Normal distribution on the right.

```
mcheck  <- function (obj,...) {
rs <-obj$resid
fv <-obj$fitted
par(mfrow=c(1,2))
plot(fv,rs,xlab="Fitted values",ylab="Residuals")
abline(h=0, lty=2)
qqnorm(rs,xlab="Normal scores",ylab="Ordered residuals")
qqline(rs,lty=2)
par(mfrow=c(1,1))
invisible(NULL)   }
```

Note the use of $ (**component selection**) to extract the residuals and fitted values from the model object which is passed to the function as obj (the expression x$name is the name **component** of x). The functions qqnorm and qqline are built-in functions to produce Normal probability plots. It is good programming practice to set the graphics parameters back to their default settings before leaving the function.

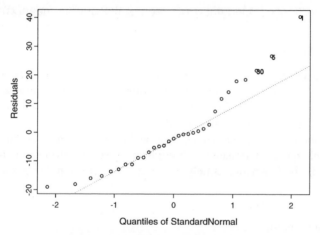

This is an example of 'banana-shaped' type of non-Normal errors (see p. 227). Other models might produce S-shaped plots of qqnorm (see p. 64).

Influence

One of the commonest reasons for a lack of fit is through the existence of outliers in the data. It is important to understand, however, that a point may **appear** to be an outlier because of mis-specification of the model, and not because there is anything wrong with the data. It is important to understand that analysis of residuals is a very poor way of looking for influence. Precisely because a point is highly influential, means that it forces the regression line close to it, and hence the influential point may have a very small residual.

Leverage

Points increase in influence to the extent that they lie on their own, a long way from the mean value of x (either left or right). To account for this, measures of leverage for a given data point y are proportional to $(x - \bar{x})^2$. The commonest measure of leverage is

$$h_i = \frac{1}{n} + \frac{(x_i - \bar{x})^2}{\sum (x_j - \bar{x})^2},$$

where the denominator is SSX (see p. 133). A good rule of thumb is that a point is highly influential if its

$$h_i > \frac{2p}{n},$$

where p is the number of parameters in the model. We could easily calculate the leverage value of each point in x. It is more efficient, perhaps, to write a general function that could carry out the calculation of $\mathbf{h}$ for any vector of x values

```
leverage < -function(x){ 1/length(x) + (x-mean(x))^2/sum((x-mean(x))^2) }.
```

Mis-specified Model

The model may have the wrong terms in it, or the terms may be included in the model in the wrong way. Here we simply note that **transformation of the explanatory variables** often produces improvements in model performance. The most frequently used transformations are logs, powers and reciprocals.

When both the error distribution and functional form of the relationship are unknown, there is no single specific rationale for choosing any given transformation in preference to another. The aim is pragmatic, namely to find a transformation that gives:

- constant error variance,
- approximately normal errors,
- additivity,
- a linear relationship between the response variables and the explanatory variables, or
- straightforward scientific interpretation.

The choice is bound to be a compromise, and as such, is best resolved by quantitative comparison of the deviance produced under different model forms. Again, in testing for non-linearity in the relationship between y and x we might add a term in x^2 to the model; a significant parameter in the x^2 term indicates curvilinearity in the relationship between y and x.

8

Regression

Regression analysis is the statistical method you use when both the response variable and the explanatory variable are continuous variables (i.e. real numbers with decimal places – things like heights, weights, volumes, or temperatures). Perhaps the easiest way of knowing when regression is the appropriate analysis is to see that a scatter plot is the appropriate graphic (in contrast to analysis of variance, say, when the plot would have been a box and whisker or a bar chart). We cover four important kinds of regression analysis:

- linear regression (the simplest, and much the most frequently used),

- polynomial regression (often used to test for non-linearity in a relationship),

- non-linear regression (to fit a specified non-linear model to data), and

- non-parametric regression (used when there is no obvious functional form).

The essence of regression analysis is using sample data to estimate parameter values and their standard errors. First, however, we need to select a model which describes the relationship between the response variable and the explanatory variable(s). The simplest model of all, is the linear model:

$$y = a + bx.$$

The response variable is y, and x is a continuous explanatory variable. There are two parameters, a and b: the intercept is a (that is the value of y when $x = 0$); and the slope is b (the slope, or gradient, is the change in y divided by the change in x which brought it about). The slope is so important, that it is worth drawing a picture to make clear what is involved.

The example refers to oil drums in a store: on week 2 there were 16 drums and when the next stock-taking was carried out on week 5 there were 10 drums left. So x is time in weeks and y is the number of full oil drums. All we know is that the graph goes through the point (2,16) and the point (5,10). Remember that when specifying coordinates on a graph (Cartesian coordinates) the x value comes first, then the y value. So the two x values

Statistics: An Introduction using R   M. J. Crawley
© 2005 John Wiley & Sons, Ltd   ISBNs: 0-470-02298-1 (PBK); 0-470-02297-3 (PPC)

are 2 and 5 and the two *y* values are 16 and 10. We see at once that *y* gets smaller as *x* increases, so the value of the slope is going to be negative. First we plot the axes of the graph, but put nothing (yet) between the axes (this is graph type = "n"):

plot(c(2,5),c(16,10),type = "n",ylab = "y",xlab = "x",ylim = c(0,20),xlim = c(0,6))

Note that in the plot function, the *x* values c(2,5) are grouped together in the first argument and the *y* values c(16,10) are grouped together in the second (i.e. the arguments of plot are **not** Cartesian coordinates even though, as here, they sometimes look as if they might be).

Let's add the two points to the graph as solid circles (this is plotting character pch = 16)

points(c(2,5),c(16,10),pch = 16)

Now to calculate the slope, we need to know the change in *y*. On the graph this is a vertical line (i.e. parallel with the *y* axis); the line representing the change in *y* would be drawn like this

lines(c(2,2),c(16,10))

Do you see how this worked? The *x* value did not change (so both *x* coordinates were 2). The top of the line was *y* = 16 and the bottom of the line was 10). Let's label this line delta *y*:

text(1,13,"delta y")

The next thing we need to calculate the slope is the change in *x*. We draw a line to represent the change in *x* like this:

lines(c(2,5),c(10,10))

We can label this line delta *x*:

text(3.5,8,"delta x")

You need to work out the *x* and *y* coordinates for locating text by trial and error after you have looked at the graph. Alternatively, there is a function called locator(1) that enables you to get the coordinate values of one point using the mouse to locate the cursor, then left-clicking. Note that text is centred on the location you specify (not, for instance, printed from a specified lower-left corner). Now, we can calculate slope, *b*, as

$$b = \frac{\text{change in } y}{\text{change in } x \text{ that brought it about}}.$$

So for our example, $b = (10 - 16)/(5 - 2) = -6/3 = -2.0$. We can draw the line with slope $= -2$ between the two points like this:

lines(c(2,5),c(16,10))

but there is a very useful function in R called abline which draws a line from exactly one edge of the plotting area to another. To use this, we need to know the value of a, the intercept. Now that we know that $b = -2.0$, this is easy. Take any one of the two known coordinates (say $\{2,16\}$) and rearrange the equation to find a. To some of you, this may be second nature, but to others it may be really hard. We'll work though this example to show what's involved. Start with what we know:

$$y = a + bx.$$

Now we know y (16), we know x (2) and we know $b(-2)$. How do we get a out of this equation? We know that $a + bx$ is equal to y, so if we subtract bx from both sides of the equation, we are left with:

$$y - bx = a + bx - bx.$$

The $+bx$ and $-bx$ cancel out, so

$$a = y - bx.$$

We can work this out for our example: $a = 16 - (-2 \times 2)$. Remember that 'minus minus equals plus' so $a = 16 + 4 = 20$. Now we are in a position to use abline to draw a line right across the plotting area: the arguments of abline are first a, then b, like this:

abline(20,-2)

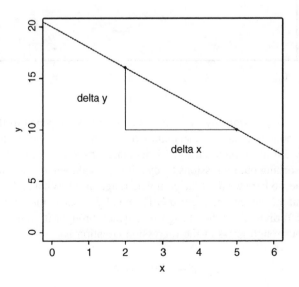

Linear Regression

Let's start with an example. The thing to understand is that there is nothing difficult, or mysterious about estimating the regression parameters. We can do a reasonable job, just by eye.

```
reg.data < -read.table("c:\\temp\\tannin.txt",header = T)
attach(reg.data)
names(reg.data)
```

```
[ 1] "growth" "tannin"
```

```
plot(tannin,growth,pch = 16)
```

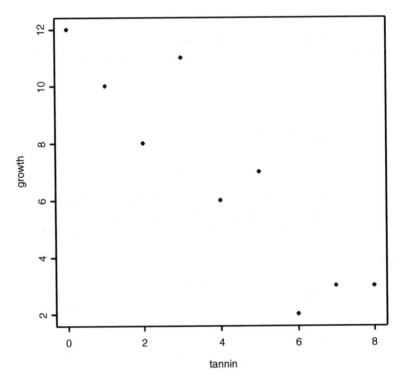

This is how we do regression 'by eye'. We ask 'what has happened to the y value ?'. It decreased from about 12 to about 2, so delta y (the change in y) $= -10$ (the minus sign is important). How did the x value change? It increased from 0 to 8, so the change in x is $+8$. (tip: when working out regressions by eye, it is a good idea to take as big a range of x values as possible, so here we took the complete range of x.). What is the value of y when $x = 0$? It is about 12, so the intercept $a = 12$. Finally, what is the value of b? It is the change in y (-10) divided by the change in x which brought it about (8), so $b = -10/8 = -1.25$. So our rough guess at the regression equation is

$$y = 12.0 - 1.25x.$$

That's all there is to it. Obviously, we want to make the procedure more objective than this. We also want to estimate the unreliability of the two estimated parameters (i.e. the standard errors of the slope and intercept), but the basics are just as straightforward as that.

Linear Regression in R

How close did we get to the maximum likelihood estimates of a and b with our guesstimates of 12 and -1.25? It is easy to find out using the R function lm which stands for 'Linear Model' (the first letter is a lower case L, not a number one). All we need do, is to tell R which of the variables is the response variable (growth in this case) and which is the explanatory variable (tannin concentration in the diet). The response variable goes on the left of the tilde ~ and the explanatory variable goes on the right: growth ~ tannin. This is read 'growth is modelled as a function of tannin'. Now we write:

lm(growth ~ tannin)

```
Coefficients:
   (Intercept)              tannin
        11.756              -1.217
```

The two parameters are called 'coefficients' in R: the intercept is 11.756 (compared with out guesstimate of 12) and the slope is -1.217 (compared with our guesstimate of -1.25). Not bad at all.

So where does R get its coefficients from? We need to do some calculations to find this out. If you are more mathematically inclined, you might like to work through Box 8.1, but this is not essential to understand what is going on. Remember that what we want are the maximum likelihood estimates of the parameters. That is to say, that given the data, and having selected a linear model, we want **to find the values of the slope and intercept that make the data most likely**. Keep re-reading this sentence until you understand what it is saying.

Box 8.1. The least-squares estimate of the regression slope, b

The **best fit** slope is found by rotating the line until the **error sum of squares**, SSE, is minimized, so we want to find the minimum of $\sum (y - a - bx)^2$. We start by finding the derivative of SSE with respect to b

$$\frac{dSSE}{db} = -2 \sum x(y - a - bx).$$

Now, multiplying through the bracketed term by x gives

$$\frac{dSSE}{db} = -2 \sum xy - ax - bx^2.$$

Apply summation to each term separately, set the derivative to zero, and divide both sides by -2 to remove the unnecessary constant:

$$\sum xy - \sum ax - \sum bx^2 = 0.$$

We cannot solve the equation as it stands because there are two unknowns, a and b. However, we know the value of a is $\bar{y} - b\bar{x}$. Also, note that $\sum ax$ can be written as $a\sum x$, so, replacing a and taking both a and b outside their summations gives:

$$\sum xy - \left[\frac{\sum y}{n} - b\frac{\sum x}{n}\right]\sum x - b\sum x^2 = 0.$$

Now multiply out the central bracketed term by $\sum x$ to get

$$\sum xy - \frac{\sum x\sum y}{n} + b\frac{(\sum x)^2}{n} - b\sum x^2 = 0.$$

Finally, take the two terms containing b to the right-hand side, and note their change of sign:

$$\sum xy - \frac{\sum x\sum y}{n} = b\sum x^2 - b\frac{(\sum x)^2}{n},$$

then divide both sides by $\sum x^2 - (\sum x)^2/n$ to obtain the required estimate b:

$$b = \frac{\sum xy - \frac{\sum x\sum y}{n}}{\sum x^2 - \frac{(\sum x)^2}{n}}.$$

Thus, the value of b that minimizes the sum of squares of the departures is given simply by

$$b = \frac{SSXY}{SSX}.$$

This is **the maximum likelihood estimate of the slope** of the linear regression.

The best way to see what is going on is to do it graphically. Let's cheat a bit by fitting the best-fit straight line through our scatterplot, using abline with a linear model, like this:

abline(lm(growth ~ tannin))

The fit is reasonably good, but it is not perfect. The data points do not lie on the fitted line. The difference between each data point and the value predicted by the model at the same value of x is called a **residual**. Some residuals are positive (above the line) and others are negative (below the line). Let's draw vertical lines to indicate the size of the residuals. The first x point is at tannin $= 0$. The y value measured at this point was growth $= 12$, but what is the growth predicted by the model at tannin $= 0$? There is a built-in function called predict to work this out:

```
fitted < -predict(lm(growth ~ tannin))
fitted
```

```
       1          2          3          4          5          6          7          8
11.755556  10.538889  9.322222   8.105556   6.888889   5.672222   4.455556   3.238889
       9
2.022222
```

So the first predicted value of growth is 11.75555. To draw the first residual, both x coordinates will be 0. The first y coordinate will be 12 (the observed value) and the second will be 11.7555 (the fitted (or predicted) value). We use lines, like this:

```
lines(c(0,0),c(12,11.7555555))
```

We could go through, laboriously, and draw each residual like this, but it is much quicker to automate the procedure, using a loop to deal with each residual in turn:

```
for (i in 1:9) lines (c(tannin[i],tannin[i]),c(growth[i],fitted[i]))
```

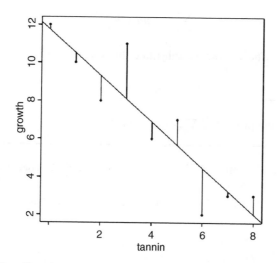

These residuals describe the goodness of fit of the regression line. Our maximum likelihood model is defined as **the model that minimizes the sum of the squares of these residuals**. It is useful, therefore, to write down exactly what any one of the residuals, d, is – it is the measured value, y, minus the fitted value called $\hat{y}$ (y 'hat'):

$$d = y - \hat{y}.$$

We can improve on this, because we know that $\hat{y}$ is on the straight line $a + bx$, so

$$d = y - (a + bx) = y - a - bx.$$

The equation includes $a - bx$ because of the minus sign outside the bracket. Now our best fit line, by definition, is given by the values of a and b that minimize the sums of the squares of the d's (see Box 8.1). Note, also, that just as $\sum(y - \bar{y}) = 0$ (Box 4.1), so $\sum d = \sum(y - a - bx) = 0$ (Box 8.2).

Box 8.2. Proof that $\sum(y - a - bx) = 0$

Take the summation through each of the terms, bearing in mind that $\sum a = na$ and $\sum bx = b\sum x$

$$\sum y - na - b\sum x.$$

We also know that the linear regression $y = a + bx$ passes through the point $(\bar{x}, \bar{y})$ defined by the mean values of x and y, so it must be the case that $\bar{y} = a + b\bar{x}$. Replacing $\bar{y}$ by $\sum y/n$ and $\bar{x}$ by $\sum x/n$ allows us to work out the value of $\sum y$

$$\frac{\sum y}{n} = a + b\frac{\sum x}{n},$$

so

$$\sum y = na + b\sum x$$

because the n's cancel. Now we substitute this value for $\sum y$:

$$\sum y - na - b\sum x = na + b\sum x - na - b\sum x = 0$$

as required for the proof that $\sum(y - a - bx) = 0$.

We want to find the minimum of $\sum d^2 = \sum(y - a - bx)^2$. To work this out we need 'the famous five' which are $\sum y^2$ and $\sum y$, $\sum x^2$ and $\sum x$ and a new quantity, $\sum xy$, the sum of products. The sum of products is worked out pointwise, so for our data, it is:

tannin

[1] 0 1 2 3 4 5 6 7 8

growth

[1] 12 10 8 11 6 7 2 3 3

tannin*growth

[1] 0 10 16 33 24 35 12 21 24

zero times $12 = 0$, plus one times $10 = 10$, plus two times $8 = 16$, and so on:

sum(tannin*growth)

[1] 175

The next thing is to use the famous five to work out three essential 'corrected sums': the corrected sum of squares of x, the corrected sum of squares of y and the corrected sum of products, $x.y$. The corrected sums of squares of x and y should be familiar to you:

$$SSY = \sum y^2 - \frac{\left(\sum y\right)^2}{n}$$

$$SSX = \sum x^2 - \frac{\left(\sum x\right)^2}{n}$$

because if we wanted the variance in y, we would just divide SSY by its degrees of freedom (and likewise for the variance in x; see p. 137). It is only the corrected sum of products that is novel, but its structure is directly analogous. Think about the formula for SSY, above. It is 'the sum of y times y' $\sum y^2$, 'minus the sum of y times the sum of y' $\left(\sum y\right)^2$ 'divided by the sample size', n. The formula for SSX is similar. It is 'the sum of x times x' $\sum x^2$, 'minus the sum of x times the sum of x' $\left(\sum x\right)^2$ 'divided by the sample size', n:

$$SSXY = \sum xy - \frac{\left(\sum x\right)\left(\sum y\right)}{n}.$$

If you look carefully you will see that the corrected sum of products has exactly the same kind of structure. It is 'the sum of x times y' $\sum xy$, 'minus the sum of x times the sum of y' $\left(\sum x\right)\left(\sum y\right)$ 'divided by the sample size', n.

These three corrected sums of squares are absolutely central to everything that follows about regression analysis, so it is a good idea to re-read this section as often as necessary, until you are confident that you understand what SSX, SSY and $SSXY$ represent (Box 8.3).

Box 8.3. Corrected sums of squares and products in regression

The total sum of squares is SSY, the corrected sum of products is $SSXY$ and the sum of squares of x is SSX:

$$SSY = \sum y^2 - \frac{\left(\sum y\right)^2}{n}$$

$$SSX = \sum x^2 - \frac{\left(\sum x\right)^2}{n}$$

$$SSXY = \sum xy - \frac{\sum x \sum y}{n}.$$

The explained variation is the regression sum of squares, *SSR*:

$$SSR = \frac{SSXY^2}{SSX}.$$

The unexplained variation is the error sum of squares, *SSE*, can be obtained by difference

$$SSE = SSY - SSR,$$

but *SSE* is defined as **the sum of the squares of the residuals** which is

$$SSE = \sum (y - a - bx)^2.$$

The correlation coefficient, *r*, is given by

$$r = \frac{SSXY}{\sqrt{SSX \times SSY}}.$$

The next question is how we use *SSX*, *SSY* and *SSXY* to find the maximum likelihood estimates of the parameters and their associated standard errors. It turns out that this step is much simpler than what has gone before. The maximum likelihood estimate of the slope, *b*, is just:

$$b = \frac{SSXY}{SSX}$$

(the detailed derivation of this is in Box 8.1). Now that we know the value of the slope, we can use any point on the fitted straight line to work out the maximum likelihood estimate of the intercept, *a*. One part of the definition of the best-fit straight line is that it passes through the point $(\bar{x}, \bar{y})$ determined by the mean values of *x* and *y*. Since we know that $y = a + bx$, it must be the case that $\bar{y} = a + b\bar{x}$, and so

$$a = \bar{y} - b\bar{x} = \frac{\sum y}{n} - b.\frac{\sum x}{n}.$$

Box 8.4. The shortcut formula for the sum of products, *SSXY*

SSXY is based on the expectation of the product $(x - \bar{x})(y - \bar{y})$. Start by multiplying out the brackets:

$$(x - \bar{x})(y - \bar{y}) = xy - x\bar{y} - y\bar{x} + \bar{x}.\bar{y}.$$

Now apply the summation remembering that $\sum \bar{y} = n.\bar{y}$ and $\sum x.\bar{y} = \bar{y}\sum x$

$$\sum xy - \bar{y}\sum x - \bar{x}\sum y + n.\bar{x}.\bar{y} = \sum xy - n.\bar{y}.\bar{x} - n.\bar{x}.\bar{y} + n.\bar{x}.\bar{y} \sum xy - n.\bar{x}.\bar{y}$$

because $\sum x = n.\bar{x}$ and $\sum y = n.\bar{y}$. Now replace the product of the two means by $\sum x/n \times \sum y/n$

$$\sum xy - n\frac{\sum x}{n}.\frac{\sum y}{n}$$

which, on cancelling the n's gives the corrected sum of products as

$$SSXY = \sum xy - \frac{\sum x \sum y}{n}.$$

We can work out the parameter values for our example. To keep things as simple as possible, we can call the variables *SSX*, *SSY* and *SSXY* (note that R is 'case sensitive' so the variable *SSX* is different from *ssx*):

```
SSX = sum(tannin^2)-sum(tannin)^2/length(tannin)
SSX
```

[1] 60

```
SSY = sum(growth^2)-sum(growth)^2/length(growth)
SSY
```

[1] 108.8889

```
SSXY = sum(tannin*growth)-sum(tannin)*sum(growth)/length(tannin)
SSXY
```

[1] -73

That's all we need. So the slope is:

$$b = \frac{SSXY}{SSX} = \frac{-73}{60} = -1.2166667$$

and the intercept is given by:

$$a = \frac{\sum y}{n} - b.\frac{\sum x}{n} = \frac{62}{9} + 1.2166667\frac{36}{9} = 6.8889 + 4.86667 = 11.755556.$$

Now we can write the maximum likelihood regression equation in full:

$$y = 11.75556 - 1.216667x.$$

This, however, is only half of the story. In addition to the parameter estimates, $a = 11.756$ and $b = -1.2167$, we need to measure the unreliability associated with each of the estimated parameters. In other words, we need to calculate the standard error of the intercept and the standard error of the slope. We have already met the standard error of a mean, and we used it in calculating confidence intervals (p. 45) and in doing Student's t-test (p. 77). Standard errors of regression parameters are similar in so far as they are enclosed inside a big square root term (so that the units of the standard error are the same as the units of the parameter), and they have the error variance, s^2, in the numerator. There are extra components, however, which are specific to the unreliability of a slope or an intercept (see Boxes 8.6 and 8.7 for details), but we cannot work out the standard errors until we know the value of the error variance s^2 and to do this, we need to carry out an analysis of variance.

Error Variance in Regression: $SSY = SSR + SSE$

The idea is simple – we take the total variation in y, SSY, and partition it into components that tell us about the explanatory power of our model. The variation that is explained by the model is called the regression sum of squares (denoted by SSR), and the unexplained variation is called the error sum of squares (denoted by SSE that we drew on the scatterplot, earlier). Then $SSY = SSR + SSE$ (the proof is presented in Box 8.5). Now, in principle, we could compute SSE because we know that it is the sum of the squares of the deviations of the data points from the fitted model, $\sum d^2 = \sum (y - a - bx)^2$. Since we know the values of a and b, we are in a position to work this out. The formula is fiddly, however, because of all those subtractions, squarings and addings-up. Fortunately, there is a very simple shortcut that involves computing SSR, the explained variation, rather than SSE. This is because

$$SSR = b.SSXY$$

so we can immediately work out $SSR = -1.21667 \times -73 = 88.81667$; and since $SSY = SSR + SSE$ we can get SSE by subtraction:

$$SSE = SSY - SSR = 108.8889 - 88.81667 = 20.07222.$$

These components are now drawn together in what is known as the 'Anova table'. Strictly, we have analysed sums of squares, rather than variances up to this point, but you will see why it is called analysis of variance shortly. The left-most column of the Anova table lists the sources of variation: regression, error and total in our example. The next column contains the sums of squares, SSR, SSE and SSY. The third column is in many ways the most important to understand; it contains the degrees of freedom. There are n points on the graph ($n = 9$ in this example). So far, our table looks like this.

Source	Sum of squares	Degrees of freedom	Mean squares	F ratio
Regression	88.817			
Error	20.072			
Total	108.889			

We shall work out the degrees of freedom associated with each of the sums of squares in turn. The easiest to deal with is the total sum of squares, because it always has the same formula for its degrees of freedom. The definition is $SSY = \sum(y - \bar{y})^2$, and you can see that there is just one parameter estimated from the data: the mean value, $\bar{y}$. Because we have estimated one parameter from the data, we have $n - 1$ degrees of freedom. The next easiest to work out is the error sum of squares. Let's look at its formula to see how many parameters need to be estimated from the data before we can work out $SSE = \sum(y - a - bx)^2$. We need to know the values of both a and b before we can calculate SSE. These are estimated from the data, so the degrees of freedom for error are $n - 2$. This is important, so re-read the last sentence if you don't see it yet. The most difficult of the three is the regression degrees of freedom, because you need to think about this one in a different way. The question is this: how many extra parameters, over and above the mean value of y, did you estimate when fitting the regression model to the data? The answer is one. The extra parameter you estimated was the slope, b. So the regression degrees of freedom in this simple model, with just one explanatory variable, is 1. This will only become clear with practice. To complete the Anova table, we need to understand the fourth column, headed 'mean squares'. This column contains the variances, on which analysis of variance is based. The key point to recall is that

$$\text{variance} = \frac{\text{sum of squares}}{\text{degrees of freedom}}.$$

This is very easy to calculate in the context of the Anova table, because the relevant sums of squares and degrees of freedom are in adjacent columns. Thus the regression variance is just $SSR/1 = SSR$, and the error variance is $s^2 = SSE/(n - 2)$. Traditionally, one does not fill in the bottom box (it would be the overall variance in y, $SSY/(n - 1)$). Finally, the Anova table is completed by working out the F ratio, which is a ratio between two variances. In most simple Anova tables, you divide the treatment variance in the numerator (the regression variance in this case) by the error variance s^2 in the denominator. The null hypothesis under test in a linear regression is that the slope of the regression line is zero (i.e. no dependence of y on x). The two-tailed alternative hypothesis is that the slope is significantly different from zero (either positive or negative). In many applications it is not particularly interesting to reject the null hypothesis, because we are interested in the estimates of the slope and its standard error (we often know from the outset that the null hypothesis is false). To test whether the F ratio is sufficiently large to reject the null hypothesis, we compare the calculated value of F in the final column of the Anova table with the critical value of F, expected by chance alone (this is found from quantiles of the F distribution qf, with 1 d.f. in the numerator and $n - 2$ d.f. in the denominator, as described below). Here is the completed Anova table:

Source	Sum of squares	Degrees of freedom	Mean squares	F ratio
Regression	88.817	1	88.817	30.974
Error	20.072	7	$s^2 = 2.86746$	
Total	108.889	8		

Notice that the component degrees of freedom add up to the total degrees of freedom (this is always true, in any Anova table, and is a good check on your understanding of the design of the experiment). The last question concerns the magnitude of the F ratio $= 30.974$: is it big enough to justify rejection of the null hypothesis? The critical value of the F ratio is the value of F that would arise due to chance alone when the null hypothesis was true, given that we have 1 d.f. in the numerator and 7 d.f. in the denominator. We have to decide on the level of uncertainty that we are willing to put up with; the traditional value for work like this is 5%, so our certainty $= 0.95$. Now we can use quantiles of the F distribution, qf, to find the critical value:

qf(0.95,1,7)

[1] 5.591448

Because our calculated value of F (30.974) is much larger than the critical value (5.591), we can be confident in rejecting the null hypothesis. Perhaps a better thing to do, rather than working rigidly at the 5% uncertainty level, is to ask what is the probability of getting a value for F as big as 30.974 or larger if the null hypothesis is true. For this we use $1 - $pf rather than qf:

1-pf(30.974,1,7)

[1] 0.0008460725

It is very unlikely indeed ($p < 0.001$). Up to this point we have assumed that $SSY = SSR + SSE$; see Box 8.5 for the proof.

Box 8.5. Proof that $SSY = SSR + SSE$

Let's start with what we know. The difference between y and $\bar{y}$ is the sum of the differences $(y - \hat{y})$ and $(\hat{y} - \bar{y})$ as you can see from the figure:

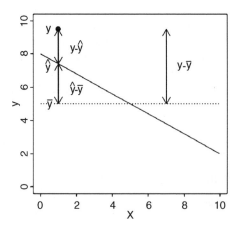

However, it is not at all obvious that the sum of the squares of these quantities, $(y - \hat{y})^2 + (\hat{y} - \bar{y})^2$, should be equal to $(y - \bar{y})^2$. We begin by squaring $(y - \hat{y}) + (\hat{y} - \bar{y})$ to see where this gets us. Remember that $(a + b)^2$ is $a^2 + b^2 + 2ab$ so let's write out the square of the sum in full:

$$(y - \hat{y})^2 + (\hat{y} - \bar{y})^2 + 2(y - \hat{y})(\hat{y} - \bar{y}).$$

Now we apply summation

$$\sum (y - \hat{y})^2 + \sum (\hat{y} - \bar{y})^2 + 2 \sum (y - \hat{y})(\hat{y} - \bar{y}).$$

The first term is $SSE = \sum (y - \hat{y})^2$, the second term is $SSR = \sum (\hat{y} - \bar{y})^2$ and the third term, $2 \sum (y - \hat{y})(\hat{y} - \bar{y})$, needs to be equal to zero if SSY is to be equal to $SSE + SSR$ as we aim to prove. The first step is to replace $\hat{y}$ and $\bar{y}$ by their relations to x in the right-most term: $\hat{y} = a + bx$ and $\bar{y} = a + b\bar{x}$

$$2 \sum (y - a - bx)(a + bx - (a + b\bar{x})).$$

The minus sign outside the inner bracket means that this becomes

$$2 \sum (y - a - bx)(bx - b\bar{x})$$

because the a's cancel out. This summation will be zero if both $\sum (y - a - bx)$ and $\sum (bx - b\bar{x})$ are zero. We have already proved that $\sum (y - a - bx) = 0$ in Box 8.2 and that $\sum (x - \bar{x}) = 0$ in Box 4.1 (admittedly in the guise of $\sum (y - \bar{y}) = 0$), so all we need to note is that $\sum (bx - b\bar{x})$ can be written as $b \sum (x - \bar{x})$ to complete the proof ($b \times 0 = 0$).

Next, we can use the calculated error variance $s^2 = 2.867$ to work out the standard errors of the slope (Box 8.6) and the intercept (Box 8.7). First the standard error of the slope:

$$\text{s.e.}_b = \sqrt{\frac{s^2}{SSX}} = \sqrt{\frac{2.867}{60}} = 0.2186.$$

Box 8.6. The standard error of the regression slope, b, is given by: $\text{s.e.}_b = \sqrt{\frac{s^2}{SSX}}$

The error variance s^2 comes from the Anova table and is the quantity used in calculating standard errors and confidence intervals for the parameters, and in carrying out hypothesis testing. SSX measures the spread of the x values along the x axis. Recall that standard errors are **unreliability estimates**. Unreliability increases with the error variance so it makes sense to have s^2 in the numerator (on top of the division). It is less

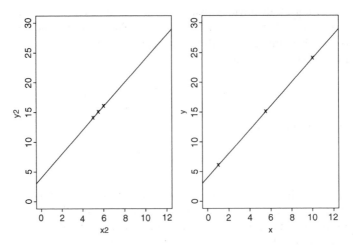

obvious why unreliability should depend on the range of x values. Look at these two graphs that have exactly the same slopes and intercepts. The difference is that the left-hand graph has all of its x values close to the mean value of x, while the graph on the right has a broad span of x values. Which of these do you think would give the most reliable estimate of the slope? It is pretty clear that it is the graph on the right, with the wider range of x values. Increasing the spread of the x values reduces unreliability of the estimated slope and hence appears in the denominator (on the bottom of the equation). What is the purpose of the big square root term? This is there to make sure that the units of the unreliability estimate are the same as the units of the parameter whose unreliability is being assessed. The error variance is in units of y **squared**, but the slope is in units of y per unit change in x.

The formula for the standard error of the intercept is a little more involved (Box 8.7)

$$s.e._{\cdot a} = \sqrt{\frac{s^2 \sum x^2}{n \times SSX}} = \sqrt{\frac{2.867 \times 204}{9 \times 60}} = 1.0408.$$

Box 8.7. The standard error of the intercept, a, is given by: $s.e._{\cdot a} = \sqrt{\dfrac{s^2 \sum x^2}{n.SSX}}$

This is like the formula for the standard error of the slope, but with two additional terms. Uncertainty declines with increasing sample size n. It is less clear why uncertainty should increase with $\sum x^2$. The reason for this is that uncertainty in the estimate of the intercept increases, the further away from the intercept that the mean value of x lies. You can see this from the following graphs. On the left is a graph with a low value of $\bar{x}$ and on the right an identical graph (same slope and intercept) but estimated from a data set with a higher value of $\bar{x}$. In both cases there is a 25% variation in the slope. Compare the difference in the prediction of the intercept in the two cases.

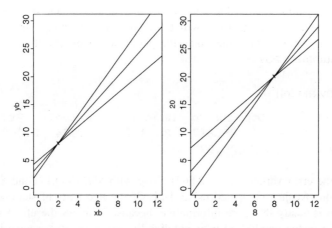

Confidence in predictions made with linear regression declines with the square of the distance between the mean value of x and the value of x at which the prediction is to be made (i.e. with $(x - \bar{x})^2$). Thus, when the origin of the graph is a long way from the mean value of x, the standard error of the intercept will be large, and vice versa. In general, the **standard error for a predicted value** $\hat{y}$ is given by:

$$\text{s.e.}_{\hat{y}} = \sqrt{s^2 \left[\frac{1}{n} + \frac{(x - \bar{x})^2}{SSX} \right]}.$$

Note that the formula for the standard error of the intercept is just the special case of this for $x = 0$ (you should check the algebra of this result as an exercise).

Now that we know where all the numbers come from, we can repeat the analysis in R and see just how straightforward it is. It is good practice to give the statistical model a name: 'model' is as good as any.

model < -lm(growth ~ tannin)

Then, you can do a variety of things with the model. The most important, perhaps, is to see the details of the estimated effects, which you get from the summary function:

summary(model)

```
Coefficients:
               Estimate    Std. Error    t value    Pr(>|t|)
(Intercept)     11.7556       1.0408      11.295    9.54e-06   ***
tannin          -1.2167       0.2186      -5.565    0.000846   ***

Residual standard error: 1.693 on 7 degrees of freedom
Multiple R-Squared: 0.8157,    Adjusted R-squared: 0.7893
F-statistic: 30.97 on 1 and 7 DF, p-value: 0.0008461
```

This shows everything you need to know about the parameters and their standard errors (compare the values for s.e.$_a$ and s.e.$_b$ with those you calculated long-hand, above). If you want to see the Anova table rather than the parameter estimates, then the appropriate function is summary.aov

summary.aov(model)

	Df	Sum Sq	Mean Sq	F value	Pr (>F)	
tannin	1	88.817	88.817	30.974	0.000846	***
Residuals	7	20.072	2.867			

This shows the error variance ($s^2 = 2.867$) along with SSR (88.817) and SSE (20.072), and the p value we just computed using 1-pf. Of the two sorts of summary table, the summary.lm is vastly the more informative, because it shows the effect sizes (in this case the slope of the graph) and their unreliability estimates (the standard error of the slope). Generally, you should resist the temptation to put Anova tables in your written work. The important information like the p-value and the error variance can be put in the text, or in figure legends, much more efficiently. Anova tables put far too much emphasis on hypothesis testing, and show nothing directly about effect sizes.

Measuring the Degree of Fit, r^2

There is a very important issue that remains to be considered. Two regression lines can have exactly the same slopes and intercepts, and yet be derived from completely different relationships as shown in the figures below. We need to be able to quantify the degree of

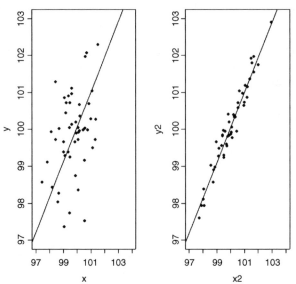

fit, which is low in the graph on the left and high in the right. In the limit, all the data points might fall exactly on the line. The degree of scatter in that case would be zero and the fit would be perfect (we might define a perfect fit as 1). At the other extreme, x might explain none of the variation in y at all; in this case, fit would be zero and the degree of scatter would be 100%. Can we combine what we have learned about SSY, SSR and SSE into a measure of fit that has these properties? Our proposed metric is **the fraction of the total variation in y that is explained by the regression**. The total variation is SSY and the explained variation is SSR, so our measure – let's call it r^2 – is given by

$$r^2 = \frac{SSR}{SSY}.$$

This varies from 1, when the regression explains all of the variation in y ($SSR = SSY$), to 0 when the regression explains none of the variation in y ($SSE = SSY$).

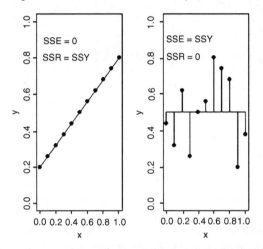

The formal name of this quantity is the coefficient of determination, but these days most people just refer to it as 'r *squared*'. We have already met the square root of this quantity (r or ρ), as the correlation coefficient (p. 93).

Model Checking

The final thing you will want to do is to expose the model to critical appraisal. The assumptions we really want to be sure about are **constancy of variance** and **normality of errors**. The simples way to do this is with four model-checking plots:

plot(model)

The first graph (top left) shows residuals on the y axis against fitted values on the x axis. It takes experience to interpret these plots, but what you ***don't*** want to see is lots of structure or pattern in the plot. Ideally, as here, the points should look like the sky at night. It is a major problem if the scatter increases as the fitted values get bigger; this would show up like a wedge of cheese on its side (see p. 200). However, in our present case, everything is all right on the constancy of variance front. The next plot (top right) shows the Normal qqnorm plot (p. 64) which should be a straight line if the errors are normally distributed. Again, the present example looks fine. If the pattern were S-shaped

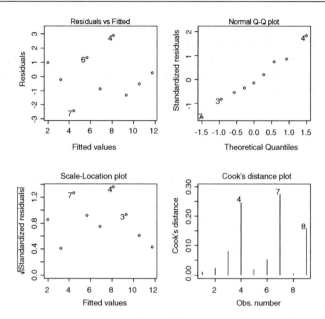

or banana-shaped, we would need to fit a different model to the data. The third plot (bottom left) is a repeat of the first, but on a different scale. It shows the square root of the standardized residuals (where all the values are positive) against the fitted values. If there was a problem, the points would be distributed inside a triangular shape, with the scatter of the residuals increasing as the fitted values increase; but there is no such pattern here, which is good. The fourth and final plot (lower right) shows Cook's distance for each of the observed values of the response variable (in the order in which they appear in the dataframe). The point of this plot is to highlight those y values that have the biggest effect on the parameter estimates (i.e. it shows **influence**; p. 123). You can see that point number 7 is the most influential; but which point is that? You can use 7 as a subscript (i.e. in square brackets) to find out:

tannin[7];growth[7]

```
[ 1]  6
[ 1]  2
```

The most influential point was the one with tannin $= 6\%$ and growth rate $= 2$. You might like to investigate how much this influential point (6,2) affected the parameter estimates and their standard errors. To do this, we repeat the statistical modelling but leave out the point in question, using subset like this (!= means 'not equal to'):

model2 < -update(model,subset = (tannin != 6))
summary(model2)

```
Coefficients:
                Estimate    Std. Error    t value     Pr (>|t|)
(Intercept)     11.6892       0.8963       13.042     1.25e-05    ***
tannin          -1.1171       0.1956       -5.712     0.00125      **
```

```
Residual standard error: 1.457 on 6 degrees of freedom
Multiple R-Squared: 0.8446,     Adjusted R-squared: 0.8188
F-statistic: 32.62 on 1 and 6 DF, p-value: 0.001247
```

First of all, notice that we have lost one degree of freedom, because there are now eight values of y rather than nine. The estimate of the slope has changed from -1.2167 to -1.1171 (a difference of about 9%) and the standard error of the slope has changed from 0.2186 to 0.1956 (a difference of about 12%). What you do in response to this information depends on the circumstances. Here, we would simply note that point (6,2) was influential and stick with our first model, using all the data. In other circumstances, a data point might be so influential that the structure of the model is changed completely by leaving it out. In that case, we might gather more data or, if the study was already finished, we might publish both results (with and without the influential point) so that the reader could make up their own mind about the interpretation. The important point is that we always do model-checking; the summary.lm(model) table is not the end of the process of regression analysis.

Polynomial Regression

The relationship between y and x often turns out not to be a straight line; but Occam's Razor requires that we fit a linear model unless a non-linear relationship is significantly better at describing the data. So this begs the question: how do we assess the significance of departures from linearity? One of the simplest ways is to use polynomial regression.

The idea of polynomial regression is straightforward. As before, we have just one continuous explanatory variable, x, but we can fit higher powers of x, like x squared and x cubed to the model in addition to x to explain curvature in the relationship between y and x. It is useful to experiment with the kinds of curves that can be generated with very simple models. Even if we restrict ourselves to the inclusion of a quadratic term, x^2, there are many curves we can describe, depending upon the signs of the linear and quadratic terms:

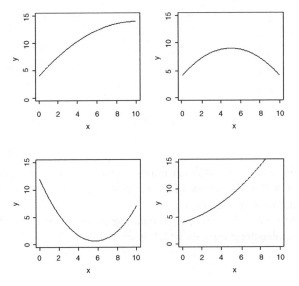

In the top left panel, there is a curve with positive but declining slope, with no hint of a hump ($y = 4 + 2x - 0.1x^2$). Top right shows a curve with a clear maximum ($y = 4 + 2x - 0.2x^2$), and bottom left shows a curve with a clear minimum ($y = 12 - 4x + 0.35x^2$). The bottom right curve shows a positive association between y and x with the slope increasing as x increases ($y = 4 + 0.5x + 0.1x^2$). So you can see that a simple quadratic model with three parameters (an intercept, a slope for x, and a slope for x^2) is capable of describing a wide range of functional relationships between y and x. It is very important to understand that the quadratic model **describes** the relationship between y and x; it does not pretend to **explain** the mechanistic (or causal) relationship between y and x.

We can see how polynomial regression works by analysing an example. Here are data showing the relationship between radioactive emissions (y) and time (x):

```
rm(x,y)
par(mfrow=c(1,1))
curve <-read.table("c:\\temp\\decay.txt",header=T)
attach(curve)
names(curve)
```

```
[1] "x" "y"
```

```
plot(x,y,pch=16)
abline(lm(y~x))
```

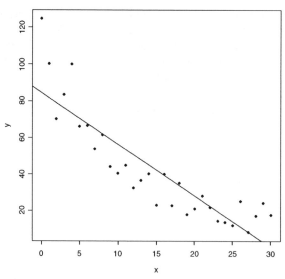

The fitted straight line (using **abline**) draws attention to the curvature. Most of the residuals for low and high values of x are positive, and most of the residuals for intermediate values of x are negative.

There are several ways of fitting polynomial regression models, but the simplest is to calculate a new explanatory variable which is x^2:

```
x2 <-x^2
```

Now we do a multiple regression with two continuous explanatory variables: x and x^2

```
quadratic < -lm(y ~ x + x2)
summary(quadratic)
```

```
Coefficients:
                Estimate    Std. Error    t value    Pr(>|t|)
(Intercept)    106.38880       4.65627     22.849     < 2e-16    ***
x               -7.34485       0.71844    -10.223    5.90e-11    ***
x2               0.15059       0.02314      6.507    4.73e-07    ***

Residual standard error: 9.205 on 28 degrees of freedom
Multiple R-Squared: 0.908,      Adjusted R-squared: 0.9014
F-statistic: 138.1 on 2 and 28 DF,   p-value: 3.109e-015
```

The equation of the model is $y = 106.3888 - 7.34485x + 0.15059x^2$, and the standard errors of each of the three parameters appear in column 3. The key point here is that the quadratic term is highly significant ($p = 4.73 \ 10^{-7}$), so there is strong evidence that the relationship is non-linear. Another way to come to the same conclusion is to use anova to compare linear and quadratic models, like this:

```
linear < -lm(y ~ x)
anova(quadratic,linear)
```

```
Analysis of Variance Table

Model 1: y ~ x + x2
Model 2: y ~ x
  Res.Df      RSS    Df    Sum of Sq        F      Pr(>F)
1     28   2372.6
2     29   5960.6    -1      -3588.1   42.344    4.727e-07    ***
```

Note that the significance of the difference is exactly the same as the significance of the quadratic term in the first model ($p = 4.73 \ 10^{-7}$). So we can conclude that there is significant non-linearity, but is the quadratic model the best description of this non-linearity? We can get some impression by plotting the fitted values from the quadratic model on our initial scatterplot. We generate a series of x values between 0 and 30 and then use these in predict with the quadratic model to generate a smooth curve of 'y hat' against x:

```
xv < -seq(0,30,0.1)
yv < -predict(quadratic,list(x = xv,x2 = xv^2))
lines(xv,yv)
```

The fit looks reasonably good, but model checking with plot(quadratic) suggests a degree of non-normality in the errors. The apparent increase in y at the highest values of x is also rather suspect (this problem often arises with polynomial models). Because the data relate to a decay process, it might be that an exponential function $y = ae^{-bx}$

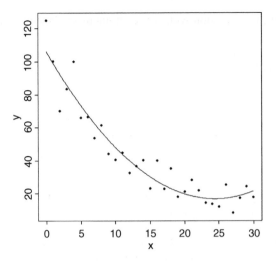

describes the data better than a quadratic. This is a question of model comparison. We can use lm to fit an exponential curve if we make ln(y) rather than y the response variable:

```
exponential < -lm(log(y) ~ x)
yv2 < -exp(predict(exponential,list(x = xv)))
lines(xv,yv2,lty = 2)
```

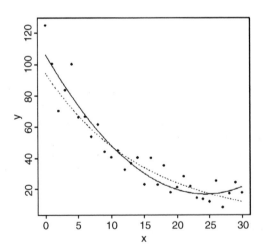

Evidently, the exponential model (dotted line, lty = 2) provides a more intuitive description of the decay process, even though plot(exponential) draws attention to quite serious non-constancy of errors. This is a good example of how insights based on a mechanistic understanding of the process (e.g. a decay curve would probably not have a minimum)

need to be weighed against statistical rules of thumb (e.g. higher r^2 is better) in model comparisons.

Non-linear Regression

Sometimes we have a mechanistic model for the relationship between y and x, and we want to estimate the parameters and standard errors of the parameters of a specific non-linear equation from data. What we mean in this case by non-linear is not that the relationship is curved (it was curved in the case of polynomial regressions, but these were linear models), but that the relationship cannot be linearized by transformation of the response variable or the explanatory variable (or both). Here is an example, which shows jaw bone length as a function of age in deer. Theory indicates that the relationship is an 'asymptotic exponential' with three parameters:

$$y = a - be^{-cx}.$$

In R, the main difference between linear models and non-linear models is that we have to tell R the exact nature of the equation as part of the model formula when we use non-linear modelling. In place of lm we write nls (this stands for 'non-linear least squares'). Then we write y ~ a-b*exp(-c*x) to spell out the precise non-linear model we want R to fit to the data. The slightly tedious thing is that R requires us to specify initial guesses at the values of the parameters a, b and c (note, however, that some common non-linear models have 'self-starting' versions in R which bypass this step; see ? nls). Let's plot the data to work out sensible starting values. It always helps in cases like this to work out the equation's 'behaviour at the limits'. That is to say, the values of y when $x = 0$ and when $x = $ infinity. For $x = 0$, we have $\exp(-0)$ which is 1, and $1 \times b = b$ so $y = a - b$. For $x = $ infinity, we have $\exp(-\text{infinity})$ which is 0, and $0 \times b = 0$ so $y = a$. That is to say, the asymptotic value of y is a, and the intercept is $a - b$.

```
deer < -read.table("c:\\temp\\jaws.txt",header = T)
attach(deer)
names(deer)
```

```
[ 1] "age" "bone"
```

```
plot(age,bone,pch = 16)
```

Inspection suggests that a reasonable estimate of the asymptote is $a \approx 120$ and intercept ≈ 10, so $b = 120 - 10 = 110$. Our guess of the value of c is slightly harder. Where the curve is rising most steeply, jaw length is about 40 where age is 5; rearranging the equation gives

$$c = -\frac{\log[(a - y)/b]}{x} = -\frac{\log[120 - 40)/110]}{5} = 0.06369075.$$

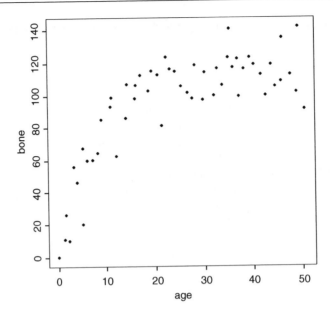

Now that we have the three parameter estimates, we can provide them to R as the starting conditions as part of the nls call like this: list(a = 120, b = 110, c = 0.064)

```
library(nls)
model < -nls(bone ~ a-b*exp(-c*age),start = list(a = 120,b = 110,c = 0.064))
summary(model)
```

```
Formula: bone~a - b * exp (-c * age)
```

```
Parameters:
     Estimate      Std. Error    t value     Pr (>|t} )
a 115.2528          2.9139        39.55       < 2e-16      ***
b 118.6875          7.8925        15.04       < 2e-16      ***
c 0.1235            0.0171         7.22       2.44e-09     ***
```

```
Residual standard error: 13.21 on 51 degrees of freedom
```

All the parameters appear to be significant at $p < 0.001$, but beware. This does not necessarily mean that all the parameters need to be retained in the model. In this case, $a = 115.2528$ with s.e. $= 2.9139$ is clearly not significantly different from $b = 118.6875$ with s.e. $= 7.8925$ (they would need to differ by more than 2 s.e. to be significant). So we should try fitting the simpler two-parameter model

$$y = a(1 - e^{-cx})$$

```
model2 < -nls(bone ~ a*(1-exp(-c*age)),start = list(a = 120,c = 0.064))
anova(model,model2)
```

```
Analysis of Variance Table
```

```
Model 1: bone ~ a - b * exp (-c * age)
Model 2: bone ~ a * (1 - exp (-c * age))
```

	Res.Df	Res.Sum Sq	Df	Sum Sq	F value	Pr (>F)
1	51	8897.3				
2	52	8929.1	-1	-31.8	0.1825	0.671

Model simplification was clearly justified ($p = 0.671$), so we accept the two-para-meter version, model2, as our minimal adequate model. We finish by plotting the curve through the scatterplot. The age variable needs to go from 0 to 50:

```
av < -seq(0,50,0.1)
```

and we use predict with model2 to generate the predicted bone lengths:

```
bv < -predict(model2,list(age = av))
lines(av,bv)
```

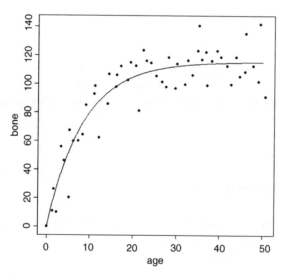

The parameters of this curve are obtained from model2:

```
summary(model2)
```

```
Parameters:
```

	Estimate	Std. Error	t value	Pr (>|t|)	
a	115.58056	2.84365	40.645	< 2e-16	***
c	0.11882	0.01233	9.635	3.69e-13	***

```
Residual standard error: 13.1 on 52 degrees of freedom
```

which we could write like this $y = 115.58(1 - e^{-0.1188x})$ or like this $y = 115.58 \left[1 - \exp(-0.1188x) \right]$ according to taste or journal style. If you want to present the standard errors as well as the parameter estimates, you could write 'the model $y = a[1 - \exp(-bx)]$ had $a = 115.58 \pm 2.84$ (1 s.e.) and $b = 0.1188 \pm 0.0123$ (1 s.e., $n = 54$) and explained 84.6% of the total variation in bone length'. Note that because there are only two parameters in the minimal adequate model, we have called them a and b (rather than a and c as in the original formulation).

Testing for Humped Relationships

Proving the existence of humps in a relationship between y and x is controversial and can be difficult, but it is easy to appreciate the issues that are involved. For instance, is there good evidence for a hump in the following relationship?

```
smooth <-read.table("c:\\temp\\smoothing.txt",header=T)
attach(smooth)
names(smooth)
```

```
[ 1] "x" "y"
```

```
par(mfrow=c(1,2))
plot(x,y)
abline(lm(y~x))
```

This is the most parsimonious relationship between y and x with just two parameters; the intercept and the slope of the linear regression. At the other extreme, we could produce a model which explained *all* of the variation in y. This is what it would look like:

```
sequence <-order(x)
plot(x,y)
lines(x[sequence],y[sequence])
```

The model has as many parameters as there are data points, hence it has no degrees of freedom, and exhibits no explanatory power. What we need is a model of intermediate complexity that optimizes the trade-off between the number of parameters and explanatory power. Incidentally, look what a mess you get if you try lines(x,y); this illustrates the advantage of using ordered subscripts. You can carry out a quadratic polynomial regression on these data as an exercise, but does the existence of a significant quadratic term in the model prove the existence of a significant hump in the relationship?

Generalized Additive Models (gams)

Sometimes we can see that the relationship between y and x is non-linear but we don't have any theory or any mechanistic model to suggest a particular functional form (mathematical equation) to describe the relationship. In such circumstances, gams are

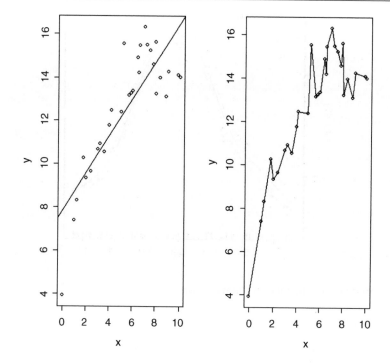

particularly useful, because they fit non-parametric smoothers to the data without requiring us to specify any particular mathematical model to describe the non-linearity. This will become clear with an example.

```
rm(x,y)
library(mgcv)
hump < -read.table("c:\\temp\\hump.txt",header = T)
attach(hump)
names(hump)
```

```
[ 1]  "y" "x"
```

We start by fitting the generalized additive model as a smoothed function of x, $s(x)$:

```
model < -gam(y ~ s(x))
```

then we plot the model, and overlay the scattergraph of data points

```
plot(model)
points(x,y-mean(y))
```

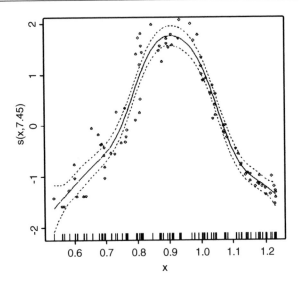

Model summary is obtained in the usual way:

summary(model)

```
Family: gaussian
Link function: identity

Formula:
y ~ s(x)

Parametric coefficients:
               Estimate    std. err.    t ratio      Pr (> |t|)
constant        1.9574      0.03446      56.8       < 2.22e-16

Approximate significance of smooth terms:
          edf     chi.sq        p-value
s(x)     7.452    982.38       < 2.22e-16

Adjusted r-sq. = 0.919       GCV score = 0.1156
Scale estimate = 0.1045              n = 88
```

This shows that the humped relationship between y and x is highly significant (see the p-value of the smooth term $s(x)$ with an r^2 of 0.919). Note that because of the strong hump in the relationship, a linear model $lm(y \sim x)$ indicates no significant relationship between the two variables ($p = 0.346$). This is an object lesson in always plotting the data before you come to conclusions from the statistical analysis; in this case, if you had started with a linear model you would have thrown out the baby with the bathwater by concluding that nothing was happening. In fact, something very significant is happening but it is producing a humped, rather than a trended relationship.

9

Analysis of Variance

Analysis of variance is the technique we use when all the explanatory variables are categorical. The explanatory variables are called **factors**, and each factor has two or more **levels**. When there is a single factor with three or more levels we use one-way Anova. If we had a single factor with just two levels, we would use Student's test (see p. 76), and this would give us exactly the same answer that we would have obtained by Anova (remember the rule that $F = t^2$). Where there are two or more factors, then we use two-way or three-way Anova, depending on the number of explanatory variables. When there is replication at each level in a multi-way Anova, the experiment is called a **factorial** design, and this allows us to study **interactions** between variables, in which we test whether the response to one factor depends on the level of another factor.

One-way Anova

There is a real paradox about analysis of variance, which often stands in the way of a clear understanding of exactly what is going on. The idea of analysis of variance is to compare two or more means, but it does this by comparing variances. How can that work?

The best way to see what is happening is to work through a graphical example. To keep things as simple as possible, we shall use a factor with just two levels at this stage, but the argument extends to any number of levels. Suppose that we have atmospheric ozone concentrations measured in parts per hundred million (pphm) in two commercial lettuce-growing gardens (we shall call the gardens A and B for simplicity).

```
oneway < -read.table("c:\\temp\\oneway.txt",header = T)
attach(oneway)
names(oneway)
```

```
[ 1]  "ozone"     "garden"
```

As usual, we begin by plotting the data, but here we plot the y values (ozone concentrations) against the order in which they were measured:

```
plot(1:20,ozone,ylim = c(0,8),ylab = "y",xlab = "order")
```

Statistics: An Introduction using R   M. J. Crawley
© 2005 John Wiley & Sons, Ltd   ISBNs: 0-470-02298-1 (PBK); 0-470-02297-3 (PPC)

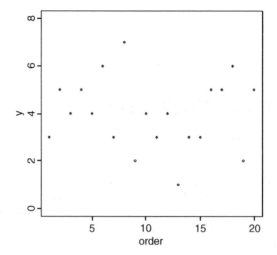

There is lots of scatter, indicating that the variance in y is large. To get a feel for the overall variance, we can plot the mean value of y and indicate each of the residuals by a line from **mean(y)** to the value of y:

```
abline(mean(ozone),0)
for(i in 1:20) lines(c(i,i),c(mean(ozone),ozone[i]))
```

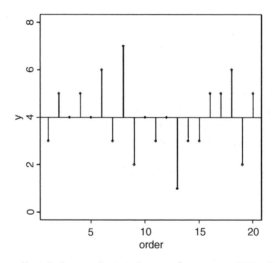

We refer to this overall variation as the **total sum of squares**, *SSY*, which more formally is given by:

$$SSY = \sum (y - \bar{y})^2$$

which should look familiar, because it is the formula used in defining the variance of y (s^2 = sum of squares/degrees of freedom; see p. 37).

This next step is the key to understanding how analysis of variance works. Instead of fitting the overall mean value of *y* through the data, and looking at the departures of all the data points from the overall mean, let's fit the individual treatment means (the mean for garden A and the mean for garden B in this case), and look at the departures of data points from the appropriate treatment mean. It will be useful if we have different plotting symbols for the different gardens; say open circles (pch = 1) for garden A and solid circles (pch = 16) for garden B. Note the type of type = "n" to suppress plotting when we first draw the axes:

```
plot(1:20,ozone,ylim = c(0,8),type = "n",ylab = "y",xlab = "order")
```

Now add the points for garden A:

```
points(seq(1,19,2),ozone[garden = = "A"],pch = 1)
```

To space out the points, we put data from the two gardens in alternating positions on the graph, using seq(1,19,2) for garden A and seq(2,20,2) for garden B:

```
points(seq(2,20,2),ozone[garden = = "B"],pch = 16)
```

Now it is clear that the mean ozone concentration in garden B is substantially higher. The aim of analysis of variance is to determine whether it is significantly higher, or whether this kind of difference could come about by chance alone, when the mean ozone concentrations in the two gardens was really the same.

Now we draw the residuals–the differences between the measured ozone concentrations, and the means for the gardens involved:

```
abline(mean(ozone[garden = = "A"]),0)
abline(mean(ozone[garden = = "B"]),0)

k < - -1
for (i in 1:10){
k < -k+2
lines(c(k,k),c(mean(ozone[garden = = "A"]),ozone[garden = = "A"] [i]))
lines(c(k+1,k+1),c(mean(ozone[garden=="B"]),ozone[garden=="B"] [i]))
}
```

This raises some questions. If the means in the two gardens are not significantly different, what should be the difference in the lengths of the residual lines in this figure and the figure before? After a bit of thought, you should see that if the means were the same, then the two horizontal lines in this figure would be in the same place, and hence the lengths of the residual lines would be the same as in the previous figure. We're half way there. Now, suppose that mean ozone concentration is different in the two gardens. Would the residual lines be bigger or smaller when we compute them from the individual treatment means (as above), or from the overall mean (as in the previous figure)? They would be **smaller** when computed from the individual treatment means **if the individual treatment means were different**.

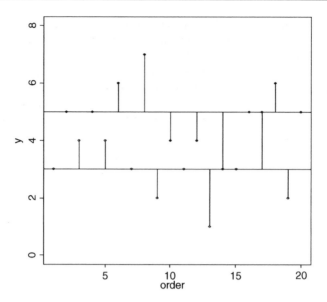

So there it is. That is how analysis of variance works. **When the means are significantly different, then the sum of squares computed from the individual treatment means will be smaller than the sum of squares computed from the overall mean.** We judge the significance of the difference between the two sums of squares using analysis of variance.

The analysis is formalized by defining this new sum of squares: it is the sum of the squares of the differences between the individual y values and the relevant treatment mean. We shall call this SSE, the **error sum of squares** (there has been no error in the sense of a mistake; 'error' is used here as a synonym of 'residual'):

$$SSE = \sum_{j=1}^{k} \sum (y - \bar{y}_j)^2.$$

We compute the mean for the jth level of the factor in advance, and then add up the squares of the differences. Given that we worked it out this way, can you see how many degrees of freedom should be associated with SSE? Suppose that there were n replicates in each treatment ($n = 10$ in our example). And suppose that there are k levels of the factor ($k = 2$ in our example). If you estimate k parameters from the data before you can work out SSE, then you must have lost k degrees of freedom in the process. Since each of the k levels of the factor has n replicates, there must be $k \times n$ numbers in the whole experiment ($2 \times 10 = 20$ in our example). So the degrees of freedom associated with SSE is $k.n - k = k(n - 1)$. Another way of seeing this is to say that there are n replicates in each treatment, and hence $n - 1$ degrees of freedom for error in each treatment (because 1 d.f. is lost in estimating each treatment mean). There are k treatments (i.e. k levels of the factor) and hence there are $k \times (n - 1)$ d.f. for error in the experiment as a whole.

Now we come to the 'analysis' part of the analysis of variance. The total sum of squares in y, SSY, is broken up (analysed) into components. The unexplained part of the variation is called the error sum of squares, *SSE*. The component of the variation that is explained by differences between the treatment means is called the treatment sum of squares, and is traditionally denoted by *SSA*. This is because in two-way analysis of variance, with two different categorical explanatory variables, we shall use *SSB* to denote the sum of squares attributable to differences between the means of the second factor, *SSC* to denote the sum of squares attributable to differences between the means of the third factor, and so on.

Analysis of variance, therefore, is based on the notion that we break down the total sum of squares, *SSY*, into useful and informative components:

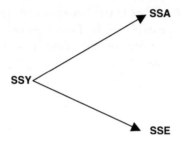

Typically, we compute all but one of the components, then find the value of the last component by subtraction of the others from *SSY*. We already have a formula for *SSE*, so we could obtain *SSA* by difference: $SSA = SSY - SSE$. Starting with *SSY* we calculate the sum of the squares of the differences between the *y* values and the overall mean:

```
SSY < -sum((ozone-mean(ozone))^2)
SSY
```

```
[ 1] 44
```

The question now is 'how much of this 44 is attributable to differences between the means of gardens A and B (*SSA* = explained variation) and how much is sampling error (*SSE* = unexplained variation)?'. We have a formula defining *SSE*; it is the sum of the squares of the residuals calculated separately for each garden, using the appropriate mean value. For garden A we get

```
sum((ozone[garden=="A"]-mean(ozone[garden=="A"]))^2)
```

```
[ 1] 12
```

and for garden B

```
sum((ozone[garden=="B"]-mean(ozone[garden=="B"]))^2)
```

```
[ 1] 12
```

so the error sum of squares is the total of these components $SSE = 12 + 12 = 24$. Finally, we can obtain the treatment sum of squares, *SSA*, by difference: $SSA = 44 - 24 = 20$.

At this point, we can fill in the Anova table (see p. 136):

Source	Sum of squares	Degrees of freedom	Mean square	F-ratio
Garden	20.0	1	20.0	15.0
Error	24.0	18	$s^2 = 1.3333$	
Total	44.0	19		

We need to test whether an *F*-ratio of 15.0 is large or small. To do this we compare it with the critical value of *F* from quantiles of the *F*-distribution, qf. We have one degree of freedom in the numerator, and 18 degrees of freedom in the denominator, and we want to work at 95% certainty ($\alpha = 0.05$):

qf(0.95,1,18)

[1] 4.413873

The calculated value of 15.0 is much greater than the critical value of $F = 4.41$, so we can reject the null hypothesis (equality of the means) and accept the alternative hypothesis (the two means are significantly different). We used a one-tailed *F*-test (0.95 rather than 0.975 in the qf function) because we are only interested in the case where the treatment variance is large relative to the error variance. This approach is rather old-fashioned; the modern view is to calculate the **effect size** (the difference between the means is 2.0 pphm ozone) and to state the probability that such a difference would arise by chance alone when the difference between the means was actually 0. For this we use cumulative probabilities of the *F* distribution, rather than quantiles, like this:

1-pf(15.0,1,18)

[1] 0.001114539

So the probability of obtaining data as extreme as ours (or more extreme) if the two means really were the same is roughly one tenth of 1%.

That was quite a lot of work. Here is the whole analysis in R in a single line:

summary(aov(ozone ~ garden))

```
           Df   Sum Sq   Mean Sq   F value      Pr (>F)
garden      1   20.0000   20.0000        15   0.001115 **
Residuals  18   24.0000    1.3333
```

The first column shows the sources of variation (*SSA* and *SSE* respectively); note that R leaves off the row that we had included for total variation, *SSY*. The next column shows degrees of freedom: there are two levels of garden (A and B) so there is $2 - 1 = 1$ d.f. for garden, and there are 10 replicates per garden, so $10 - 1 = 9$ d.f. per garden and two gardens, so error d.f. $= 2 \times 9 = 18$. The next column shows the sums of squares: $SSA = 20$ and $SSE = 24$. The fourth column gives the mean squares (sums of squares divided by degrees of freedom); the treatment mean square is 20.0 and the error variance, s^2 (synonym of the residual mean square) is 1.3333. The F ratio is 15, and the probability that this (or a more extreme result) would arise by chance alone if the two means really were the same, is 0.001115 (as we calculated long-hand, above).

We finish by carrying out graphical checks of the assumptions of the model, namely constancy of variance and normality of errors.

```
plot(aov(ozone ~ garden))
```

The first plot on your screen shows that the variances are identical in the two treatments (this is exactly what we want to see). The second plot shows a reasonably straight-line relationship on the Normal quantile–quantile plot (especially since, in this example, the y values are whole numbers), so we can be confident that non-normality of errors is not a major problem. The third plot shows the residuals against the fitted values on a different scale, and the fourth plot shows Cook's statistics, drawing attention to the fact that points 8, 13 and 14 are potentially influential. We can test for their influence by repeating the analysis but leaving out these points:

```
wanted = (1:20 != 8 & 1:20 != 13 & 1:20 != 14)
wanted
```

```
[ 1] TRUE TRUE TRUE TRUE TRUE TRUE TRUE FALSE TRUE TRUE TRUE TRUE
[13] FALSE FALSE TRUE TRUE TRUE TRUE TRUE TRUE
```

We can use subset to leave out the three potentially influential points:

```
summary(aov(ozone ~ garden,subset = wanted))
```

	Df	Sum Sq	Mean Sq	F value	Pr (>F)
garden	1	13.3856	13.3856	17.376	0.0008239 ***
Residuals	15	11.5556	0.7704		

The interpretation is unaffected; only the degrees of freedom (15 instead of 18 d.f. for error) and the p value have changed.

Shortcut Formula

In the unlikely event that you ever need to do analysis of variance using a calculator, then it is useful to know the shortcut formula for calculating SSA. We calculated it by difference, above, having worked out SSE longhand. To do this, the thing you need to understand is what we mean by a 'treatment total'. The treatment total is simply the sum of the y values in a particular factor level. For our two gardens we have:

cbind(ozone[garden=="A"],ozone[garden=="B"])

```
        [ ,1] [ ,2]
 [ 1,]    3     5
 [ 2,]    4     5
 [ 3,]    4     6
 [ 4,]    3     7
 [ 5,]    2     4
 [ 6,]    3     4
 [ 7,]    1     3
 [ 8,]    3     5
 [ 9,]    5     6
 [ 10,]   2     5
```

tapply(ozone,garden,sum)

```
 A        B
30       50
```

The totals for gardens A and B are 30 and 50 respectively, and we shall call these T_1 and T_2. The shortcut formula for SSA (Box 9.1) is then:

$$SSA = \frac{\sum T_i^2}{n} - \frac{(\sum y)^2}{kn}.$$

We should check that this really does give the correct value for SSA:

$$SSA = \frac{30^2 + 50^2}{10} - \frac{80^2}{2 \times 10} = \frac{3400}{10} - \frac{6400}{20} = 340 - 320 = 20$$

which checks out. In all sorts of analysis of variance, the key point to realize is that the sum of the subtotals squared is always **divided by the number of numbers that were added together to get each subtotal**. That sounds complicated, but the idea is simple. In our case we squared the subtotals T_1 and T_2 and added the results together. We divided by 10 because both T_1 and T_2 were the sum of ten numbers.

Box 9.1. Corrected sums of squares in one-way Anova

The definition of the total sum of squares, SSY, is the sum of the squares of the differences between the data points, y, and the overall mean, $\bar{\bar{y}}$

$$SSY = \sum_{i=1}^{k} \sum (y - \bar{\bar{y}})^2$$

---

where $\sum$ means the sum over the n replicates within each of the k factor levels. the error sum of squares, SSE, is the sum of the squares of the differences between the data points, y, and their individual treatment means, $\bar{y}_i$

$$SSE = \sum_{i=1}^{k} \sum (y - \bar{y}_i)^2.$$

The treatment sum of squares, SSA, is the sum of the squares of the differences between the individual treatment means, $\bar{y}_i$, and the overall mean, $\bar{\bar{y}}$

$$SSA = \sum_{i=1}^{k} \sum_{j=1}^{n} (\bar{y}_i - \bar{\bar{y}})^2 = n \sum_{i=1}^{k} (\bar{y}_i - \bar{\bar{y}})^2.$$

Squaring the bracketed term, and applying summation gives

$$\sum \bar{y}_i^2 - 2\bar{\bar{y}} \sum \bar{y}_i + k.\bar{\bar{y}}^2.$$

Now replace $\bar{y}_i$ by T_i/n (where T is our conventional name for the k individual treatment totals) and replace $\bar{\bar{y}}$ by $\sum y/k.n$ to get

$$\frac{\sum_{i=1}^{k} T_i^2}{n^2} - 2\frac{\sum y \sum_{i=1}^{k} T_i}{n.k.n} + k\frac{\sum y \sum y}{k.n.k.n}.$$

Note that $\sum_{i=1}^{k} T_i = \sum_{i=1}^{j} \sum_{j=1}^{n} y_{ij}$ so the right-hand positive and negative terms both have the form $(\sum y)^2/k.n^2$. Finally, multiplying through by n gives

$$SSA = \frac{\sum T^2}{n} - \frac{(\sum y)^2}{k.n}.$$

As an exercise, you should prove that $SSY = SSA + SSE$ (and see Box 8.5).

Effect Sizes

So far we have concentrated on hypothesis testing, using summary.aov. It is usually more informative to investigate the effects of the different factor levels, using summary.lm like this:

summary.lm(aov(ozone ~ garden))

It was easy to interpret this kind of output in the context of a regression, where the rows represent parameters that are intuitive–namely, the intercept and the slope. In the context of analysis of variance, it takes a fair bit of practice before the meaning of this kind of output is transparent.

```
Coefficients:
                Estimate    Std. Error    t value    Pr(>|t|)
(Intercept)       3.0000        0.3651      8.216    1.67e-07 ***
gardenB           2.0000        0.5164      3.873    0.00111 **

Residual standard error:  1.155 on 18 degrees of freedom
Multiple R-Squared: 0.4545,      Adjusted R-squared:  0.4242
F-statistic:            15 on 1 and 18 DF,     p-value: 0.001115
```

The rows are labelled (Intercept) and gardenB, but what do the parameter estimates 3.0 and 2.0 actually mean? Why are the standard errors different in the two rows? After all, the variances in the two gardens were identical.

To understand the answers to these questions, we need to know how the equation for the explanatory variables is structured when the explanatory variable, as here, is categorical. To recap, the linear regression model is written as

$$lm(y \sim x)$$

which R interprets as the two-parameter linear equation

$$y = a + bx$$

in which the values of the parameters a and b are to be estimated from the data. But what about our analysis of variance? We have one explanatory variable, $x =$ 'garden', with two levels, 'A' and 'B'. The aov model is exactly analogous to the regression model

$$aov(y \sim x)$$

but what is the associated equation? Let's look at the equation first, then try to understand it:

$$y = a + bx_1 + cx_2.$$

This looks just like a multiple regression, with two explanatory variables, x_1 and x_2. The key point to understand is that x_1 and x_2 are **the levels of the factor called x**. If 'garden' was a four-level factor, then the equation would have four explanatory variables in it, x_1 to x_4. With a categorical explanatory variable, the levels are all coded as 0 except for the level associated with the y value in question, which is coded as 1. You will find this hard to understand without a good deal of practice. Let's look at the first row of data in our dataframe:

garden[1]

[1] A

So the first ozone value in the dataframe comes from garden A. This means that $x_1 = 1$ and $x_2 = 0$. The equation for the first row therefore looks like this:

$$y = a + b \times 1 + c \times 0 = a + b \times 1 = a + b.$$

What about the second row of the dataframe?

garden[2]

[1] B

Because this row refers to garden B, x_1 is coded as 0 and x_2 is coded as 1 so the equation becomes

$$y = a + b \times 0 + c \times 1 = a + c \times 1 = a + c.$$

So what does this tell us about the parameters a, b and c? And why do we have three parameters, when the experiment generates only two means? These are the crucial questions for understanding the summary.lm output from an analysis of variance. The simplest interpretation of the three-parameter case that we have dealt with so far is that the (Intercept) a is the overall mean from the experiment:

mean(ozone)

[1] 4

Then, if a is the overall mean, so $a + b$ must be the mean for garden A and $a+c$ must be the mean for garden B (see the equations, above). If that is true, then b must be **the difference between the mean of garden A and the overall mean**. And c must be **the difference between the mean of garden B and the overall mean**. Thus, the (Intercept) is a **mean**, and the other parameters are **differences between means**. This explains why the standard errors are different in the different rows of the table: the standard error of the intercept is the standard error of a mean

$$\text{s.e.}_{\bar{y}} = \sqrt{\frac{s_A^2}{n_A}},$$

whereas the standard errors on the other rows are standard errors of the difference between two means:

$$\text{s.e.}_{\text{diff}} = \sqrt{\frac{s_A^2}{n_A} + \frac{s_B^2}{n_B}}$$

which is a bigger number (bigger by a factor of $1.4142 = \sqrt{2}$ if, as here, the sample sizes and variances are equal).

With three parameters, then, we should have $b =$ mean ozone concentration in garden A − 4 and $c =$ mean ozone concentration in garden B − 4.

mean(ozone[garden == "A"])-mean(ozone)

```
[1] -1
```

mean(ozone[garden == "B"])-mean(ozone)

```
[1]  1
```

That would be a perfectly reasonable way to parameterize the model for this analysis of variance, but it suffers from the fact that there is a redundant parameter. The experiment produces only two means (one for each garden), and so there is no point in having three parameters to represent the output of the experiment. One of the three parameters is said to be 'aliased'. There are lots of ways round this dilemma, as explained in detail in Chapter 12 on Contrasts. Here we adopt the convention that is used as the default in R: so called **treatment contrasts**. Under this convention, we dispense with the overall mean, a. So now we are left with the right number of parameters (b and c). In treatment contrasts, **the factor level that comes first in the alphabet is set equal to the Intercept**. The other parameters are expressed as differences between this mean and the other relevant means. So, in our case, the mean of garden A becomes the intercept

mean(ozone[garden = = "A"])

```
[1]  3
```

and the difference between the means of garden B and garden A is the second parameter:

mean(ozone[garden == "B"])-mean(ozone[garden == "A"])

```
[1] 2
```

Let's revisit our summary.lm table and see if it now makes sense:

```
Coefficients:
              Estimate   Std. Error   t value   Pr(>|t|)
(Intercept)     3.0000       0.3651     8.216   1.67e-07   ***
gardenB         2.0000       0.5164     3.873   0.00111    **
```

The (Intercept) is 3.0 which is the mean for garden A (because the factor level 'A' comes before level 'B' in the alphabet). The estimate for garden B is 2.0. This tells us that the mean ozone concentration in garden B is 2 p.p.h.m. greater than in garden A (greater because there is no minus sign–absence of a sign implies 'plus'). We would compute the mean for garden B as $3.0 + 2.0 = 5.0$. In practice, we would not obtain the means like this, but by using tapply, instead:

tapply(ozone, garden, mean)

```
A  B
3  5
```

There is more about these issues in Chapter 12.

Plots for Interpreting One-way Anova

There are two traditional ways of plotting the results of Anova:

- box and whisker plots, and
- bar plots with error bars.

Here is an example to compare the two approaches. We have an experiment on plant competition with one factor and five levels. The factor is called clipping and the levels are control (i.e. unclipped) with two intensities of shoot pruning and two intensities of root pruning:

```
comp < -read.table("c:\\temp\\competition.txt",header = T)
attach(comp)
names(comp)
```

```
[ 1] "biomass" "clipping"
```

```
plot(clipping,biomass,xlab = "Competition treatment",ylab = "Biomass")
```

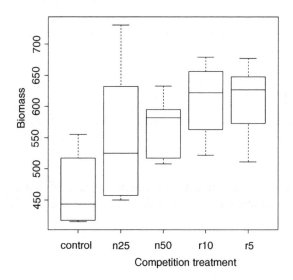

The box and whisker plot is good at showing the nature of the variation within each treatment, and also whether there is skew within each treatment (e.g. for the control plots, there is a wider range of values in the 50%–75% quartile than in the 25%–50% quartile). No outliers are shown above the whiskers, so the tops and bottoms of the bars are the maxima and minima within each treatment. The medians for the competition treatments are all higher than the 75% percentile of the controls, suggesting that they may be significantly different from the controls, but there is little to suggest that any of the competition treatments are significantly different from one another (see below for the analysis). We could use the notch = T option to get a visual impression of the significance of differences between the means; all the treatment medians fall outside the notch of the controls, but no other comparisons appear to be significant.

Barplots with error bars are preferred by many journal editors, and some people think that they make hypothesis testing easier. We shall see. Unlike S-Plus, R does not have a built-in function called error.bar so we shall have to write our own. Here is a very simple version without any bells or whistles. We shall call it error.bars to distinguish it from the much more general S-Plus function.

```
error.bars < -function(yv,z,nn){
xv < -
barplot(yv,ylim = c(0,(max(yv) + max(z))),names = nn,
    ylab = deparse(substitute(yv)))
g<- (max(xv)-min(xv))/50
for (i in 1:length(xv)) {
lines(c(xv[i],xv[i]),c(yv[i] + z[i],yv[i]-z[i]))
lines(c(xv[i]-g,xv[i] + g),c(yv[i] + z[i], yv[i] + z[i]))
lines(c(xv[i]-g,xv[i] + g),c(yv[i]-z[i], yv[i]-z[i]))
}}
```

To use this function we need to decide what kind of values (z) to use for the lengths of the bars. Let's use the standard error of a mean based on the pooled error variance from the Anova, then return to a discussion of the pros and cons of different kinds of error bars later. Here is the one-way analysis of variance:

```
model < -aov(biomass ~ clipping)
summary(model)
```

```
          Df   Sum Sq   Mean Sq   F value    Pr (>F)
clipping   4    85356    21339    4.3015   0.008752    **
Residuals 25   124020     4961
```

From the Anova table we learn that the pooled error variance $s^2 = 4961.0$. Now we need to know how many numbers were used in the calculation of each of the five means:

```
table(clipping)
```

```
clipping
control   n25   n50   r10     r5
     6     6     6     6      6
```

There was equal replication (which makes life easier), and each mean was based on six replicates, so the standard error of a mean is $\sqrt{s^2/n} = \sqrt{4961/6} = 28.75$. We shall draw an error bar up 28.75 from each mean and down by the same distance, so we need five values for z, one for each bar, each of 28.75:

```
se < -rep(28.75,5)
```

We need to provide labels for the five different bars – the factor levels should be good for this:

```
labels < -as.character(levels(clipping))
```

Now we work out the five mean values which will be the heights of the bars, and save them as a vector called ybar:

```
ybar < -as.vector(tapply(biomass,clipping,mean))
```

Now we can create the barplot with error bars:

```
error.bars(ybar,se,labels)
```

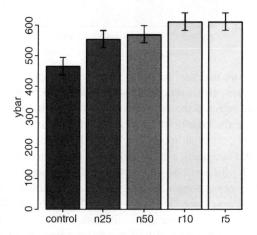

We do not get the same feel for the distribution of the values within each treatment as was obtained by the box and whisker plot, but we can certainly see clearly which means are not significantly different. If, as here, we use ±1 s.e. as the length of the error bars, then **when the bars overlap this implies that the two means are not significantly different**. Remember the rule of thumb for t: significance requires two or more standard errors, and if the bars overlap it means that the difference between the means is less than two standard errors. There is another issue, too. For comparing means, we should use the standard error of the difference between two means (not the standard error of one mean) in our tests (see p. 165); these bars would be about 1.4 times as long as the bars we have drawn here. So while we can be sure that the two root-pruning treatments are not significantly different from one another, and that the two shoot-pruning treatments are not significantly different from one another (because their bars overlap), we cannot conclude from this plot that the controls have significantly lower biomass than the rest (because the error bars are not the correct length for testing differences between means).

An alternative graphical method is to use 95% confidence intervals for the lengths of the bars, rather than standard errors of means. This is easy to do: we multiply our standard errors by Student's t qt(.975,5) $= 2.570582$ to get the lengths of the confidence intervals:

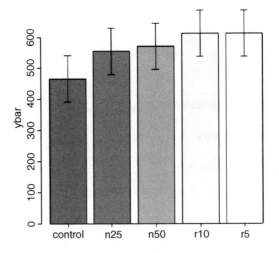

Now, all of the error bars overlap, implying visually that there are no significant differences between the means. However, we know that this is not true from our analysis of variance, in which we rejected the null hypothesis that all the means were the same at $p = 0.00875$. If it were the case that the bars did not overlap when we are using confidence intervals (as here), then that would imply that the means differed by more than four standard errors, and this is a much greater difference than is required to conclude that the means are significantly different. So this is not perfect either. With standard errors we could be sure than the means were not significantly different when the bars did overlap; and with confidence intervals we can be sure that the means are significantly different when the bars do not overlap – but the alternative cases are not clear cut for either type of bar. Can we somehow get the best of both worlds, so that the means are significantly different when the bars do not overlap, and the means are not significantly different when the bars do overlap?

The answer is yes, we can, if we use LSD bars (LSD stands for 'least significant difference'). Let's revisit the formula for Student's t-test:

$$t = \frac{\text{a difference}}{\text{standard error of the difference}}$$

and we say that the difference is significant when $t > 2$ (by the rule of thumb, or $t > $ qt(0.975,df) if we want to be more precise). We can rearrange this formula to find the smallest difference that we would regard as being significant. We can call this the least significant difference:

$$\text{LSD} = \text{qt}(0.975, \text{df}) \times \text{standard error of a difference} \approx 2 \times \text{s.e.}_{\text{difference}}.$$

In our present example this is

qt(0.975,10)*sqrt(2*4961/6)

[1] 90.60794

because a difference is based on $12 - 2 = 10$ degrees of freedom. What we are saying is the two means would be significantly different if they differed by 90.61 or more. How can we show this graphically? We want overlapping bars to indicate a difference less than 90.61, and non-overlapping bars to represent a difference greater than 90.61. With a bit of thought you will realize that we need to draw bars that are LSD/2 in length, up and down from each mean. Let's try it with our current example:

lsd < -qt(0.975,10)*sqrt(2*4961/6)
lsdbars < -rep(lsd,5)/2
error.bars(ybar,lsdbars,labels)

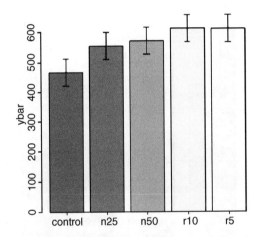

Now we can interpret the significant differences visually. The control biomass is significantly lower than any of the four treatments, but none of the four treatments is significantly different from any other. The statistical analysis of this contrast is explained in detail in Chapter 12. Sadly, most journal editors insist on error bars of 1 s.e.. It is true that there are complicating issues to do with LSD bars (not least the vexed question of multiple comparisons; see p. 226), but at least LSD/2 bars do what was intended by the error plot (i.e. overlapping bars means non-significance and non-overlapping bars means significance); neither standard errors nor confidence intervals can say that. A better option might be to use box and whisker plots with the notch = T option to indicate significance (see p. 77).

Factorial Experiments

A factorial experiment has two or more factors, each with two or more levels, plus replication for each combination of factor levels. This means that we can investigate statistical interactions, in which the response to one factor depends on the level of another factor. Our example comes from a farm-scale trial of animal diets. There are two factors: diet and supplement. Diet is a factor with three levels: barley, oats and wheat. Supplement is a factor with four levels: agrimore, control, supergain and supersupp. The response variable is weight gain after 6 weeks.

```
weights < -read.table("c:\\temp\\growth.txt",header = T)
attach(weights)
```

Data inspection is carried out using barplot (note the use of beside = T to get the bars in adjacent clusters rather than vertical stacks):

```
barplot(tapply(gain,list(diet,supplement),mean),beside = T)
```

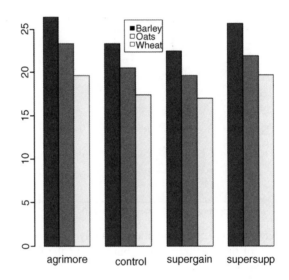

Note that the second factor in the list (supplement) appears as groups of bars from left to right in alphabetical order by factor level, from 'agrimore' to 'supersupp'. The second factor (diet) appears as three levels within each group of bars: on your screen red = barley, orange = oats, yellow = wheat, again in alphabetical order by factor level. We should really add a key to explain the levels of diet (I used locator(1) to find the appropriate coordinates for the legend at (6.3,26); this is the **top left** corner of the box):

```
labs < -c("Barley","Oats","Wheat")
cols < -c("red","orange","yellow")
legend(6.3,26,labs,fill = cols)
```

We inspect the mean values using tapply as usual:

```
tapply(gain,list(diet,supplement),mean)
```

	agrimore	control	supergain	supersupp
barley	26.34848	23.29665	22.46612	25.57530
oats	23.29838	20.49366	19.66300	21.86023
wheat	19.63907	17.40552	17.01243	19.66834

Now we use aov or lm to fit a factorial Anova (the choice affects whether we get an Anova table or a list of parameter estimates as the default output from summary). We estimate parameters for the main effects of each level of diet and each level of supplement, plus terms for the interaction between diet and supplement. Interaction degrees of freedom are the product of the degrees of freedom of the component terms, i.e. $(3 - 1) \times (4 - 1) = 6$. The model is gain $\sim$ diet + supplement + diet:supplement, but this can be simplified using the asterisk notation like this:

```
model < -aov(gain ~ diet*supplement)
summary(model)
```

	Df	Sum Sq	Mean Sq	F value	Pr (>F)	
diet	2	287.171	143.586	83.5201	2.998e-14	***
supplement	3	91.881	30.627	17.8150	2.952e-07	***
diet: supplement	6	3.406	0.568	0.3302	0.9166	
Residuals	36	61.890	1.719			

The Anova table shows that there is no hint of any interaction between the two explanatory variables ($p = 0.9166$); evidently the effects of diet and supplement are additive. The disadvantage of the Anova table is that it does not show us the effect sizes, and does not allow us to work out how many levels of each of the two factors are significantly different. As a preliminary to model simplification, summary.lm is often more useful than summary.aov:

```
summary.lm(model)
```

Coefficients:

	Estimate	Std. Error	t value	Pr (>|t|)	
(Intercept)	26.3485	0.6556	40.191	< 2e-16	***
dietoats	-3.0501	0.9271	-3.290	0.002248	**
dietwheat	-6.7094	0.9271	-7.237	1.61e-08	***
supplementcontrol	-3.0518	0.9271	-3.292	0.002237	**
supplementsupergain	-3.8824	0.9271	-4.187	0.000174	***
supplementsupersupp	-0.7732	0.9271	-0.834	0.409816	
dietoats:supplementcontrol	0.2471	1.3112	0.188	0.851571	
dietwheat:supplementcontrol	0.8183	1.3112	0.624	0.536512	
dietoats:supplementsupergain	0.2470	1.3112	0.188	0.851652	
dietwheat:supplementsupergain	1.2557	1.3112	0.958	0.344601	
dietoats:supplementsupersupp	-0.6650	1.3112	-0.507	0.615135	

```
dietwheat:supplementsupersupp          0.8024    1.3112    0.612      0.544381
```

```
Residual standard error: 1.311 on 36 degrees of freedom
Multiple R-Squared: 0.8607,   Adjusted R-squared: 0.8182
F-statistic: 20.22 on 11 and 36 DF,   p-value: 3.295e-012
```

This is a rather complex model, because there are 12 estimated parameters (the number of rows in the table), six main effects and six interactions. The output re-emphasizes that none of the interaction terms is significant, but it suggests that the minimal adequate model will require five parameters: an intercept, a difference due to oats, a difference due to wheat, a difference due to control and a difference due to supergain (these are the five rows with significance stars). This draws attention to the main shortcoming of using treatment contrasts as the default. If you look carefully at the table, you will see that the effect sizes of two of the supplements, control and supergain, are not significantly different from one another. You need lots of practice at doing t-tests in your head, to be able to do this quickly. Ignoring the signs (because the signs are negative for both of them) we have 3.05 vs. 3.99, a difference of 0.94. But look at the associated standard errors (both 0.927); the difference is only about 1 s.e. of a difference between two means. For significance, we would need roughly 2 s.e.'s (remember the rule of thumb, in which $t \geq 2$ is significant; see p. 68). The rows get starred in the significance column because treatments contrasts compare all the main effects in the rows with the intercept (where each factor is set to its first level in the alphabet, namely agrimore and barley in this case). When, as here, several factor levels are different from the intercept, but not different from one another, they all get significance stars. This means that you cannot count up the number of rows with stars in order to determine the number of significantly different factor levels. We first simplify the model by leaving out the interaction terms:

model < -aov(gain ~ diet + supplement)
summary.lm(model)

```
Coefficients:
                     Estimate   Std. Error   t value    Pr(>|t|)
(Intercept)          26.1230      0.4408      59.258      <2e-16    ***
dietoats             -3.0928      0.4408      -7.016    1.38e-08    ***
dietwheat            -5.9903      0.4408     -13.589      <2e-16    ***
supplementcontrol    -2.6967      0.5090      -5.298    4.03e-06    ***
supplementsupergain  -3.3815      0.5090      -6.643    4.72e-08    ***
supplementsupersupp  -0.7274      0.5090      -1.429       0.160
```

It is clear that we need to retain all three levels of diet (oats differ from wheat by $5.99 - 3.10 = 2.89$ with a standard error of 0.44). It is not clear that we need four levels of supplement, however. Supersupp is not obviously different from the agrimore (0.727 with s.e. $= 0.509$). Nor is supergain obviously different from the un-supplemented control animals ($3.38 - 2.70 = 0.68$). We shall try a new two-level factor to replace the four-level supplement, and see if this significantly reduces the model's explanatory power. Agrimore and supersupp are re-coded as 'best' and control and supergain as 'worst':

```
supp2 < -factor(supplement)
levels(supp2)
```

```
[ 1] "agrimore"  "control"  "supergain"  "supersupp"
```

```
levels(supp2)[c(1,4)] < -"best"
levels(supp2)[c(2,3)] < -"worst"
levels(supp2)
```

```
[ 1] "best"  "worst"
```

Now we can compare the two models

```
model2 < -aov(gain ~ diet + supp2)
anova(model,model2)
```

```
Analysis of Variance Table
```

```
Model 1: gain ~ diet + supplement
Model 2: gain ~ diet + supp2
```

```
Res.Df   RSS Df   Sum of Sq      F Pr (>F)
1   42    65.296
2   44    71.284 -2 -5.988  1.9257    0.1584
```

The simpler model two has saved two degrees and is not significantly worse than the more complex model ($p = 0.158$). This is the minimal adequate model – all of the parameters are significantly different from zero and from one another:

```
summary.lm(model2)
```

```
Coefficients:
               Estimate   Std. Error   t value    Pr (>|t|)
(Intercept)    25.7593     0.3674      70.106      <2e-16     ***
dietoats       -3.0928     0.4500      -6.873     1.76e-08    ***
dietwheat      -5.9903     0.4500     -13.311      <2e-16     ***
supp2worst     -2.6754     0.3674      -7.281     4.43e-09    ***
```

```
Residual standard error: 1.273 on 44 degrees of freedom
Multiple R-Squared: 0.8396,        Adjusted R-squared: 0.8286
F-statistic: 76.76 on 3 and 44 DF, p-value:             0
```

Model simplification has reduced our initial 12-parameter model to a four-parameter model.

Pseudoreplication: Nested Designs and Split Plots

The model-fitting functions **aov** and **lme** have the facility to deal with complicated error structures. Detailed analysis of these topics is beyond the scope of this book (see

Statistical Computing, Crawley 2002, for worked examples), but it is important that you can recognize them, and hence avoid the pitfalls of pseudoreplication. There are two general cases:

- nested sampling, as when repeated measurements are taken from the same individual, or observational studies are conduced at several different spatial scales (mostly random effects), and

- split-plot analysis, as when designed experiments have different treatments applied to plots of different sizes (mostly fixed effects)

Split-plot Experiments

In a split-plot experiment, different treatments are applied to plots of different sizes. Each different plot size is associated with its own error variance, so instead of having one error variance (as in all the Anova tables up to this point), we have as many error terms as there are different plot sizes. The analysis is presented as a series of component Anova tables, one for each plot size, in a hierarchy from the largest plot size with the lowest replication at the top, down to the smallest plot size with the greatest replication at the bottom.

The example refers to a designed field experiment on crop yield with three treatments: irrigation (with two levels, irrigated or not), sowing density (with three levels, low, medium and high), and fertilizer application (with three levels, low, medium and high).

```
yields < -read.table("c:\\temp\\splityield.txt",header = T)
attach(yields)
names(yields)
```

```
[ 1] "yield"  "block"  "irrigation"  "density"  "fertilizer"
```

The largest plots were the four whole fields (block), each of which was split in half, and irrigation was allocated at random to one half of the field. Each irrigation plot was split into three, and one of three different seed-sowing densities (low, medium or high) was allocated at random (independently for each level of irrigation and each block). Finally, each density plot was divided into three and one of three fertilizer nutrient treatments (N, P, or N and P together) was allocated at random. The model formula is specified as a factorial, using the asterisk notation. The error structure is defined in the Error() term, with the plot sizes listed from left to right, from largest to smallest, with each variable separated by the slash operator /. Note that the smallest plot size, fertilizer, does not need to appear in the error term:

```
model < -aov(yield~irrigation*density*fertilizer + Error(block/irrigation/density))
summary(model)
```

Error: block

	Df	Sum Sq	Mean Sq	F value	Pr (>F)
Residuals	3	194.444	64.815		

Error: block:irrigation

	Df	Sum Sq	Mean Sq	F value	Pr (>F)	
irrigation	1	8277.6	8277.6	17.590	0.02473	*
Residuals	3	1411.8	470.6			

Error: block:irrigation:density

	Df	Sum Sq	Mean Sq	F value	Pr (>F)	
density	2	1758.36	879.18	3.7842	0.05318.	
irrigation: density	2	2747.03	1373.51	5.9119	0.01633	*
Residuals	12	2787.94	232.33			

Error: Within

	Df	Sum Sq	Mean Sq	F value	Pr (>F)	
fertilizer	2	1977.44	988.72	11.4493	0.0001418	***
irrigation: fertilizer	2	953.44	476.72	5.5204	0.0081078	**
density: fertilizer	4	304.89	76.22	0.8826	0.4840526	
irrigation: density: fertilizer	4	234.72	58.68	0.6795	0.6106672	
Residuals	36	3108.83	86.36			

Here, you see the four Anova tables, one for each plot size: blocks are the biggest plots, half blocks get the irrigation treatment, one third of each half block gets a sowing density treatment, and one third of a sowing density treatment gets each fertilizer treatment. Note that the non-significant main effect for density ($p = 0.053$) does **not** mean that density is unimportant, because density appears in a significant interaction with irrigation (the density terms cancel out, when averaged over the two irrigation treatments; see below). The best way to understand the two significant interaction terms is to plot them using interaction.plot like this

interaction.plot(fertilizer,irrigation,yield)

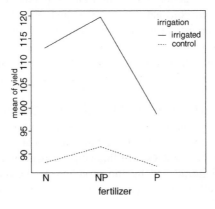

Irrigation increases yield proportionately more on the N-fertilized plots than on the P-fertilized plots. The irrigation/density interaction is more complicated:

interaction.plot(density,irrigation,yield)

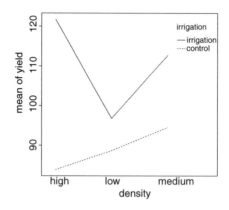

On the irrigated plots, yield is minimal on the low-density plots, but on control plots yield is minimal on the high-density plots.

Random Effects and Nested Designs

Mixed effects models are so called because the explanatory variables are a mixture of fixed effects and random effects:

- fixed effects influence only the **mean** of y,

- random effects influence only the **variance** of y.

A random effect should be thought of as coming from a population of effects: the existence of this population is an extra assumption. We speak of **prediction of random effects**, rather than estimation; we **estimate** fixed effects from data, but we intend to make predictions about the population from which our random effects were sampled. Fixed effects are unknown constants to be estimated from the data. Random effects govern the variance–covariance structure of the response variable. The fixed effects are often experimental treatments that were applied under our direction, and the random effects are either categorical or continuous variables that are distinguished by the fact that we are typically not interested in the parameter values, but only in the variance they explain.

One or more of the explanatory variables represents **grouping** in time or in space. Random effects that come from the same group will be correlated, and this contravenes one of the fundamental assumptions of standard statistical models: **independence of errors**. Mixed effects models take care of this non-independence of errors by modelling the covariance structure introduced by the grouping of the data. A major benefit of random effects models is that they economize on the number of degrees of freedom used up by the factor levels. Instead of estimating a mean for every single factor level, the random effects model estimates the distribution of the means (usually as the standard deviation of the differences of the factor-level means around an overall mean). Mixed effects models are particularly useful in cases where there is temporal pseudoreplication (repeated measurements) and/or spatial pseudoreplication (e.g. nested designs or split-plot experiments). These models can allow for:

- spatial autocorrelation between neighbours,

- temporal autocorrelation across repeated measures on the same individuals,

- differences in the mean response between blocks in a field experiment, and

- differences between subjects in a medical trial involving repeated measures.

The point is that we really do not want to waste precious degrees of freedom in estimating parameters for each of the separate levels of the categorical random variables. On the other hand, we do want to make use of the all measurements we have taken, but because of the pseudoreplication we want to take account of both the

- correlation structure, used to model within-group correlation associated with temporal and spatial dependencies, using **correlation**, and

- variance function, used to model non-constant variance in the within-group errors using **weights**.

Fixed or Random Effects?

It is difficult without lots of experience to know when to use categorical explanatory variables as fixed effects and when as random effects. Some guidelines are given below.

- Am I interested in the effect sizes ? Yes, means fixed effects.

- Is it reasonable to suppose that the factor levels come from a population of levels? Yes, means random effects.

- Are there enough levels of the factor in the data frame on which to base an estimate of the variance of the population of effects? No, means fixed effects.

- Are the factor levels informative? Yes, means fixed effects.

- Are the factor levels just numeric labels ? Yes, means random effects.

- Am I mostly interested in making inferences about the distribution of effects, based on the random sample of effects represented in the dataframe? Yes, means random effects.

- Is there hierarchical structure? Yes, means you need to ask whether the data are experimental or observations.

- Is it an hierarchical experiment, where the factor levels are experimental manipulations? Yes, means fixed effects in a split-plot design (see p. 176).

- Is it an hierarchical observational study? Yes, means random effects, perhaps in a variance components analysis (see p. 181).

- When your model contains both fixed and random effects, use mixed effects models.

- If your model structure is linear, use linear mixed effects, **lme**.

- Otherwise, specify the model equation and use non-linear mixed effects, **nlme**.

Removing the Pseudoreplication

The extreme response to pseudoreplication in a data set is simply to eliminate it. Spatial pseudoreplication can be averaged away and temporal pseudoreplication can be dealt with by carrying out separate Anovas, one at each time. This approach has two major weaknesses:

- it cannot address questions about treatment effects that relate to the longitudinal development of the mean response profiles (e.g. differences in growth rates between successive times);

- inferences made with each of the separate analyses are not independent, and it is not always clear how they should be combined.

Analysis of Longitudinal Data

The key feature of longitudinal data is that the same individuals are measured repeatedly through time. This would represent temporal pseudoreplication if the data were used uncritically in regression or Anova. The set of observations on one individual subject will tend to be positively correlated and this correlation needs to be taken into account in carrying out the analysis. The alternative is a cross-sectional study, with all the data gathered at a single point in time, in which each individual contributes a single data point. The advantage of longitudinal studies is that they are capable of separating **age effects** from **cohort effects**; these are inextricably confounded in cross-sectional studies. This is particularly important when differences between years mean that cohorts originating at different times experience different conditions, so that individuals of the same age in different cohorts would be expected to differ. There are two extreme cases in longitudinal studies:

- a few measurements on a large number of individuals,

- a large number of measurements on a few individuals,

In the first case it is difficult to fit an accurate model for change within individuals, but treatment effects are likely to be tested effectively. In the second case, it is possible to get an accurate model of the way that individuals change though time, but there is less power for testing the significance of treatment effects, especially if variation from individual to individual is large. In the first case, less attention will be paid to estimating the correlation structure, while in the second case the covariance model will be the principal focus of attention. The aims are:

- to estimate the average time course of a process,

- to characterize the degree of heterogeneity from individual to individual in the rate of the process,

- to identify the factors associated with both of these, including possible cohort effects.

The response is not the individual measurement, but the **sequence of measurements** on an individual subject. This enables us to distinguish between age effects and year effects (see Diggle *et al.* (1994) for details).

Derived Variable Analysis

The idea here is to get rid of the pseudoreplication by reducing the repeated measures into a set of summary statistics (slopes, intercepts or means), then **analyse these summary statistics** using standard parametric techniques like Anova or regression. The technique is weak when the values of the explanatory variables change through time. Derived variable analysis makes most sense when it is based on the parameters of scientifically interpretable non-linear models from each time sequence. However, the best model from a theoretical perspective may not be the best model from the statistical point of view.

There are three qualitatively different sources of random variation:

- **random effects**: experimental units differ (e.g. genotype, history, size, physiological condition) so that there are intrinsically high responders and other low responders,

- **serial correlation**: there may be time-varying stochastic variation within a unit (e.g. market forces, physiology, ecological succession, immunity) so that correlation depends on the time separation of pairs of measurements on the same individual, with correlation weakening with the passage of time,

- **measurement error**: the assay technique may introduce an element of correlation (e.g. shared bioassay of closely spaced samples; different assay of later specimens).

Variance Components Analysis (VCA)

For random effects we are often more interested in the question of how much of the variation in the response variable can be attributed to a given factor, than we are in estimating means or assessing the significance of differences between means. This procedure is called variance components analysis.

```
rats < -read.table("c:\\temp\\rats.txt",header = T)
attach(rats)
names(rats)
```

```
[ 1] "Glycogen"  "Treatment"  "Rat"  "Liver"
```

This classic example of pseudoreplication comes from Snedecor and Cochran's *Statistical Methods* (1980). Three experimental treatments were administered to rats, and the glycogen contents of the rats' livers were analysed as the response variable. This was the set-up – there were two rats per treatment, so the total sample was $n = 3 \times 2 = 6$. The tricky bit was that after each rat was killed, its liver was cut up into three pieces: a left-hand bit, a central bit and a right-hand bit. So now there are six rats each producing three bits of liver, for a total of $6 \times 3 = 18$ numbers. Finally, two separate preparations were made from each macerated bit of liver, to assess the measurement error associated with the analytical machinery. At this point there are $2 \times 18 = 36$ numbers in the dataframe as a whole. The factor levels are numbers, so we need to declare the explanatory variables to be categorical before we begin:

```
Treatment < -factor(Treatment)
Rat < -factor(Rat)
Liver < -factor(Liver)
```

Here is the analysis done the **wrong** way:

```
model < -aov(Glycogen ~ Treatment)
summary(model)
```

	Df	Sum Sq	Mean Sq	F value	Pr (>F)
Treatment	2	1557.56	778.78	14.498	3.031e-05 ***
Residuals	33	1772.67	53.72		

Treatment has a highly significant effect on liver glycogen content ($p = 0.00003$). This is wrong! We have committed a classic error of pseudoreplication. Look at the error line in the Anova table: it says the residuals have 33 degrees of freedom. However, there were only six rats in the whole experiment, so the error d.f. has to be $6 - 1 - 2 = 3$ (not 33)! Here is the analysis of variance done properly, averaging away the pseudoreplication:

```
tt < -as.numeric(Treatment)
yv < -tapply(Glycogen,list(Treatment,Rat),mean)
tv < -tapply(tt,list(Treatment,Rat),mean)

model < -aov(as.vector(yv) ~ factor(as.vector(tv)))
summary(model)
```

	Df	Sum Sq	Mean Sq	F value	Pr (>F)
factor(as.vector(tv))	2	259.593	129.796	2.929	0.1971
Residuals	3	132.944	44.315		

Now the error degrees of freedom are correct (d.f. $= 3$, not 33), and the interpretation is completely different: there are no significant differences in liver glycogen under the three experimental treatments ($p = 0.1971$).

There are two different ways of doing the analysis properly in R: Anova with multiple error terms (aov) or linear mixed effects models (lme). The problem is that the bits of the same liver are pseudoreplicates because they are spatially correlated (they come from the same rat); they are not independent, as required if they are to be true replicates. Likewise, the two preparations from each liver bit are very highly correlated (the livers were macerated before the preparations were taken, so they are essentially the same sample (certainly not independent replicates of the experimental treatments).

Here is the correct analysis using aov with multiple error terms. In the error term we start with the largest scale (treatment), then rats within treatments, then liver bits within rats within treatments. Finally, there were replicated measurements (two preparations) made for each bit of liver.

model2 < -aov(Glycogen ~ Treatment + Error(Treatment/Rat/Liver))
summary(model2)

```
Error: Treatment
                Df      Sum Sq      Mean Sq
Treatment        2     1557.56      778.78
```

```
Error: Treatment:Rat
                Df      Sum Sq      Mean Sq    F value    Pr (>F)
Residuals        3      797.67      265.89
```

```
Error: Treatment:Rat:Liver
                Df      Sum Sq      Mean Sq    F value    Pr (>F)
Residuals       12      594.0        49.5
```

```
Error: Within
                Df      Sum Sq      Mean Sq    F value    Pr (>F)
Residuals       18      381.00       21.17
```

You can do the correct, non-pseudoreplicated analysis of variance from this output (Box 9.2).

Box 9.2. Sums of squares in hierarchical designs

The trick to understanding these sums of squares is to appreciate that with nested categorical explanatory variables (random effects) the correction factor, which is subtracted from the sum of squared subtotals, is **not** the conventional $(\sum y)^2/kn$. Instead, the correction factor is the uncorrected sum of squared subtotals from the level in the hierarchy immediately above the level in question. This is very hard to see without lots of practice. The total sum of squares, *SSY*, and the treatment sum of squares, *SSA*, are computed in the usual way (see Box 9.1):

$$SSY = \sum y^2 - \frac{\left(\sum y\right)^2}{n}$$

$$SSA = \frac{\sum_{i=1}^{k} C_i^2}{n} - \frac{\left(\sum y\right)^2}{kn}.$$

The analysis is easiest to understand in the context of an example. For the rats data, the treatment totals were based on 12 numbers (two rats, three liver bits per rat and two preparations per liver bit). In this case, in the formula for *SSA* above, $n = 12$ and $kn = 36$. We need to calculate sums of squares for rats within treatments, SS_{Rats}, liver bits within rats within treatments, $SS_{Liver\ bits}$, and preparations within liver bits within rats within treatments, $SS_{Preparations}$:

$$SS_{Rats} = \frac{\sum R^2}{6} - \frac{\sum C^2}{12}$$

$$SS_{Liverbits} = \frac{\sum L^2}{2} - \frac{\sum R^2}{6}$$

$$SS_{Preparations} = \frac{\sum y^2}{1} - \frac{\sum L^2}{2}.$$

The correction factor at any level is **the uncorrected sum of squares from the level above**. The last sum of squares could have been computed by difference:

$$SS_{Preparations} = SSY - SSA - SS_{Rats} - SS_{Liverbits}.$$

The F test for equality of the treatment means is the treatment variance divided by the 'rats within treatment variance' from the row immediately beneath: $F = 778.78/265.89 = 2.928956$, with 2 d.f. in the numerator and 3 d.f. in the denominator (as we obtained in the correct Anova, above).

To turn this into a variance components analysis we need to do a little work. The mean squares are converted into variance components like this:

$$\text{Residuals} = \text{preparations within liver bits}: \text{ unchanged} = 21.17$$

$$\text{Liver bits within rats within treatments}: (49.5 - 21.17)/2 = 14.165$$

$$\text{Rats within treatments}: (265.89 - 49.5)/6 = 36.065$$

You divide the difference in variance by the number of numbers in the level below (i.e. two preparations per liver bit, and six preparations per rat, in this case).

What is the Difference Between Split-plot and Hierarchical Samples?

Split-plot experiments have informative factor levels. Hierarchical samples have uninformative factor levels. That's the distinction. In the irrigation experiment, the factor levels were as follows:

levels(density)

```
[ 1] "high" "low" "medium"
```

levels(fertilizer)

```
[ 1] "N" "NP" "P"
```

They show the density of seed sown, and the kind of fertilizer applied – they are informative. Here are the factor levels from the rats experiment:

levels(Rat)

```
[ 1] "1" "2"
```

levels(Liver)

```
[ 1] "1"   "2"   "3"
```

These factor levels are uninformative, because rat number 2 in treatment 1 has nothing in common with rat number 2 in treatment 2, or with rat number 2 in treatment 3. Liver bit number 3 from rat 1 has nothing in common with liver bit number 3 from rat 2. Note, however, that numbered factor levels are **not** always uninformative: treatment levels 1, 2 and 3 are informative: 1 is the control, 2 is a diet supplement and 3 is a combination of two supplements.

When the factor levels are informative, the variable is known as a **fixed effect**. When the factor levels are uninformative, the variable is known as a **random effect**. Generally, we are interested in fixed effects as they influence the mean, and in random effects as they influence the variance. We tend not to speak of effect-sizes attributable to random effects, but effect-sizes and their standard errors are often the principal focus when we have fixed effects. Thus, irrigation, density and fertilizer are fixed effects, and rat and liver-bit are random effects.

10

Analysis of Covariance

Analysis of covariance involves a combination of regression and analysis of variance. The response variable is continuous, and there is at least one continuous explanatory variable and at least one categorical explanatory variable. Typically, the maximal model involves estimating a slope and an intercept (the regression part of the exercise) for each level of the categorical variable(s) (the Anova part of the exercise). Let's take a concrete example. Suppose we are modelling weight (the response variable) as a function of gender and age. Gender is a factor with two levels (male and female) and age is a continuous variable. The maximal model therefore has four parameters: two slopes (a slope for males and a slope for females) and two intercepts (one for males and one for females) like this:

$$weight_{\text{male}} = a_{\text{male}} + b_{\text{male}} \times age$$
$$weight_{\text{female}} = a_{\text{female}} + b_{\text{female}} \times age.$$

Model simplification is an essential part of analysis of covariance, because the principle of parsimony requires that we keep as few parameters in the model as possible.

There are six possible models in this case, and the process of model simplification begins by asking whether we need all four parameters (top left). Perhaps we could make do with two intercepts and a common slope (top right). Or a common intercept and two different slopes (centre left). There again, age may have no significant effect on the response, so we may only need two parameters to describe the main effects of gender on weight; this would show up as two separated, horizontal lines in the plot (one mean weight for each gender; centre right). Alternatively, there may be no effect of gender at all, in which case we only need two parameters (one slope and one intercept) to describe the effect of age on weight (bottom left). In the limit, neither the continuous nor the categorical explanatory variables might have any significant effect on the response, in which case, model simplification will lead to the one-parameter null model $\hat{y} = \bar{y}$ (a single, horizontal line – bottom right).

Decisions about model simplification are based on the explanatory power of the model: if the simpler model does not explain significantly less of the variation in the response, then the simpler model is preferred. Tests of explanatory power are carried out using

Statistics: An Introduction using R M. J. Crawley
© 2005 John Wiley & Sons, Ltd ISBNs: 0-470-02298-1 (PBK); 0-470-02297-3 (PPC)

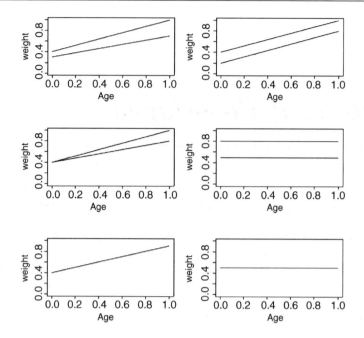

anova to compare two models: we only retain the more complicated model if the *p* value from the Anova comparing the two models is less than 0.05.

Let's see how this all works by investigating a realistic example. The dataframe concerns an experiment on a plant's ability to regrow and produce seeds following grazing. The initial, pre-grazing size of the plant is recorded as the diameter of the top of its rootstock. Grazing is a two-levels factor: grazed or ungrazed (protected by fences). The response is the weight of seeds produced per plant at the end of the growing season. Our expectation is that big plants will produce more seeds than small plants and that grazed plants will produce fewer seeds than ungrazed plants. Let's see what actually happened:

```
compensation < -read.table("c:\\temp\\compensation.txt",header = T)
attach(compensation)
names(compensation)
```

```
[ 1] "Root"    "Fruit"    "Grazing"
```

We begin with data inspection. First, did initial plant size matter?

Yes it did. Plants which were bigger to begin with produced more seeds at the end of the growing season. What about grazing?

```
plot(Grazing,Fruit)
```

This is not at all what we expected to see. Apparently, the grazed plants produced **more** seeds, not less than the ungrazed plants. We shall return to this after we have carried out

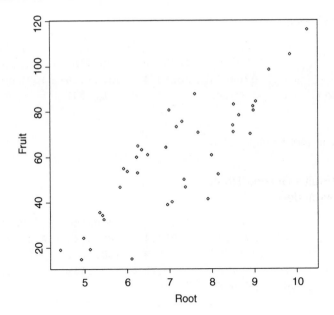

the statistical modelling. Analysis of covariance is done in the familiar way – it is just that the explanatory variables are a mixture of continuous and categorical variables. We start by fitting the most complicated model, with different slopes and intercepts for the grazed and ungrazed plants. For this, we use the asterisk operator:

model < -lm(Fruit ~ Root*Grazing)

An important thing to realize about analysis of covariance is that '**order matters**'. Look at the regression sum of squares in the Anova table when we fit root first:

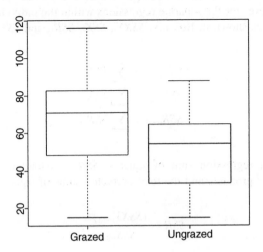

summary.aov(model)

	Df	Sum Sq	Mean Sq	F value	Pr (>F)	
Root	1	16795.0	16795.0	359.9681	<2.2e-16	***
Grazing	1	5264.4	5264.4	112.8316	1.209e-12	***
Root:Grazing	1	4.8	4.8	0.1031	0.75	
Residuals	36	1679.6	46.7			

and when we fit root second:

model < -lm(Fruit ~ Grazing*Root)
summary.aov(model)

	Df	Sum Sq	Mean Sq	F value	Pr (>F)	
Grazing	1	2910.4	2910.4	62.3795	2.262e-09	***
Root	1	19148.9	19148.9	410.4201	<2.2e-16	***
Grazing:Root	1	4.8	4.8	0.1031	0.75	
Residuals	36	1679.6	46.7			

In both cases, the error sum of squares (1679.6) and the interaction sum of squares (4.8) are the same, but the regression sum of squares (labelled 'root') is much greater when root is fitted to the model after grazing (19 148.9), than when it is fitted first (16 795.0). This is because the data for Ancova are typically non-orthogonal. Remember, **with non-orthogonal data, order matters** (Box 10.1).

Box 10.1. Corrected sums of squares in analysis of covariance

The total sum of squares, SSY, and the treatment sums of squares, SSA, are calculated in the same way as in a straightforward analysis of variance (Box 9.1). The sums of squares for the separate regressions within the individual factor levels, i, are calculated as shown in Box x.x: $SSXY_i$, SSX_i, SSR_i, and SSE_i.

$$SSXY_{total} = \sum SSXY_i$$
$$SSX_{total} = \sum SSX_i$$
$$SSR_{total} = \sum SSR_i.$$

Then the overall regression sum of squares, SSR, is calculated from the total corrected sums of products and the total corrected sums of squares of x:

$$SSR = \frac{(SSXY_{total})^2}{SSX_{total}}.$$

The difference in the two estimates, SSR and SSR_{total} is called $SSR_{difference}$ and is a measure of the significance of the differences between the regression slopes. Now we can compute SSE by difference:

$$SSE = SSY - SSA - SSR - SSR_{difference},$$

but SSE is defined for the k levels in which the regressions were computed as

$$SSE = \sum_{i=1}^{k} \sum (y - a_i - b_i x)^2.$$

Back to the analysis. The interaction, $SSR_{difference}$ representing differences in slope between the grazed and ungrazed treatments, appears to be insignificant, so we remove it:

model2 < -lm(Fruit ~ Grazing + Root)

Notice the use of + rather than * in the model formula. This says 'fit different intercepts for grazed and ungrazed plants, but fit the same slope to both graphs'. Does this simpler model have significantly lower explanatory power? We use Anova to find out:

anova(model,model2)

```
Analysis of Variance Table

Model 1: Fruit ~ Grazing + Root + Grazing:Root
Model 2: Fruit ~ Grazing + Root
  Res.Df      RSS   Df   Sum of Sq        F    Pr(>F)
1     36  1679.65
2     37  1684.46   -1       -4.81   0.1031      0.75
```

The simpler model does not have significantly lower explanatory power ($p = 0.75$), so we adopt it. Note that we did not have to do the anova in this case: the p value given in the summary.aov(model) table gave the correct, deletion p value. Here are the parameter estimates from our minimal adequate model:

summary.lm(model2)

```
Coefficients:
                 Estimate  Std. Error   t value   Pr(>|t})
(Intercept)      -127.829       9.664    -13.23   1.33e-15   ***
GrazingUngrazed    36.103       3.357     10.75   6.11e-13   ***
Root               23.560       1.149     20.51    <2e-16    ***

Residual standard error: 6.747 on 37 degrees of freedom
Multiple R-Squared: 0.9291,      Adjusted R-squared: 0.9252
F-statistic: 242.3 on 2 and 37 DF, p-value:           0
```

The model has high explanatory power, accounting for more than 90% of the variation in seed production (multiple r^2). The hard thing about analysis of covariance is understanding what the parameter estimates mean. Starting at the top, the first row, as labelled, contains an intercept. It is the intercept for the graph of seed production against initial rootstock size for the grazing treatment **whose factor level comes first in the alphabet**. To see which one this is, we can use levels:

```
levels(Grazing)
```

```
[ 1] "Grazed"    "Ungrazed"
```

So the intercept is the intercept for the grazed plants. The second row, labelled 'GrazingUngrazed' is **a difference between two intercepts**. To get the intercept for the ungrazed plants, we need to add 36.103 to the intercept for the grazed plants $(-127.829 + 36.103 = -91.726)$. The third row, labelled Root, is **a slope**: it is the gradient of the graph of seed production against initial rootstock size, and it is the same for both grazed and ungrazed plants. If there had been a significant interaction term, this would have appeared in row four as **a difference between two slopes**.

We can now plot the fitted model through the scatterplot. It will be useful to have different plotting symbols for the grazed and ungrazed plants, and the function called split comes into its own in such cases:

```
sf <-split(Fruit,Grazing)
sr <-split(Root,Grazing)
plot(Root,Fruit,type = "n",ylab = "Seed production",xlab = "Initial root
    diameter")
points(sr[[1]],sf[[1]],pch = 16)
points(sr[[2]],sf[[2]])
```

The double-bracketed subscripts on *sr* and *sf* are used because these two objects are lists rather than vectors. They are lists into which the points relating to the two grazing levels were separated by the split function.

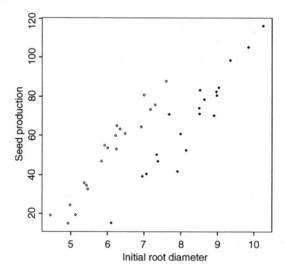

With this plot, it becomes clear why we got the curious result at the beginning (in which grazing appeared to increase seed production). The truth is that the majority of big plants ended up in the grazed treatment (the solid symbols). If you compare like with like (e.g. plants at 7mm initial root diameter) it is clear that the ungrazed plants (open symbols) produced more seed than the grazed plants (36.103 more, to be precise). This will become clearer when we fit the lines predicted by model2:

abline(−127.829,23.56)
abline(−127.829 + 36.103,23.56,lty = 2)

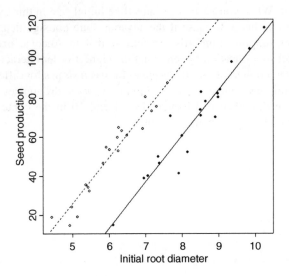

This example shows the great strength of analysis of covariance. By controlling for initial plant size, we have completely reversed the interpretation. The naïve first impression was that grazing increased seed production:

tapply(Fruit,Grazing,mean)

```
 Grazed  Ungrazed
67.9405   50.8805
```

and this was significant if we were rash enough to fit grazing on its own ($p = 0.027$):

summary(aov(Fruit ~ Grazing))

```
           Df   Sum Sq  Mean Sq  F value   Pr(>F)
Grazing     1   2910.4   2910.4   5.3086  0.02678  *
Residuals  38  20833.4    548.2
```

However, when we do the correct analysis of covariance, we find the opposite result: grazing significantly **reduces** seed production for plants of comparable initial size, e.g. from 77.46 to 41.36 at mean rootstock size:

−127.829 + 36.103 + 23.56*mean(Root)

```
[ 1] 77.4619
```

−127.829 + 23.56*mean(Root)

```
[ 1] 41.35889
```

The moral is clear. When you have covariates (like initial size in this example), then use them. This can do no harm, because if the covariates are not significant, they will drop out during model simplification. Also remember that in Ancova, **order matters**. So always start model simplification by removing the highest-order interaction terms first. In Ancova, these interaction terms are **differences between slopes** for different factor levels (recall that in multi-way Anova, the interaction terms were differences between means). Other Ancovas are described in Chapters 13, 14 and 16 in the context of count data, proportion data and binary response variables.

11

Multiple Regression

In multiple regression we have a continuous response variable and two or more continuous explanatory variables (i.e. no categorical explanatory variables). There are several important issues involved in carrying out a multiple regression:

- which explanatory variables to include,
- curvature in the response to the explanatory variables,
- interactions between explanatory variables,
- correlation between explanatory variables,
- the risk of over-parameterization.

The approach recommended here is that before you begin modelling in earnest you do two things:

- use tree models to investigate whether there are complicated interactions, and
- use generalized additive models (gam's) to investigate curvature.

A Simple Example

Let's begin with an example from air pollution studies. How is ozone concentration related to wind speed, air temperature and the intensity of solar radiation?

```
ozone.pollution < -read.table("c:\\temp\\ozone.data.txt",header = T)
attach(ozone.pollution)
names(ozone.pollution)
```
```
[ 1] "rad" "temp" "wind" "ozone"
```

In multiple regression, it is always a good idea to use pairs to look at all the correlations:

```
pairs(ozone.pollution,panel = panel.smooth)
```

Statistics: An Introduction using R   M. J. Crawley
© 2005 John Wiley & Sons, Ltd   ISBNs: 0-470-02298-1 (PBK); 0-470-02297-3 (PPC)

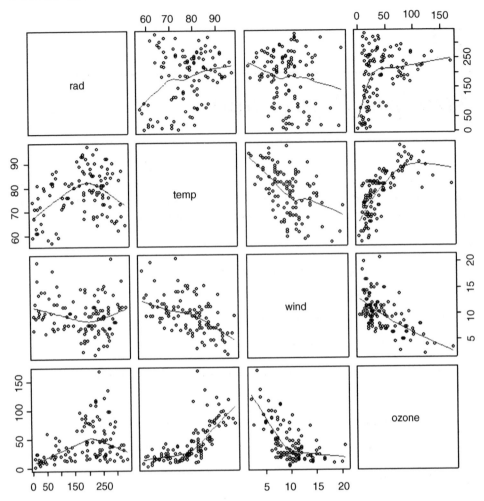

The response variable, ozone concentration, is shown on the y axis of the bottom row of panels: there is a strong negative relationship with wind speed, a positive correlation with temperature and a rather unclear, but possibly humped relationship with radiation.

A good way to start a multiple regression problem is using non-parametric smoothers in a generalized additive model (gam) like this:

```
library(mgcv)
par(mfrow = c(2,2))
model < -gam(ozone ~ s(rad) + s(temp) + s(wind))
plot(model)
par(mfrow = c(1,1))
```

The confidence intervals are sufficiently narrow to suggest that the curvature in the relationship between ozone and temperature is real, but the curvature of the relationship with wind is questionable, and a linear model may well be all that is required for solar

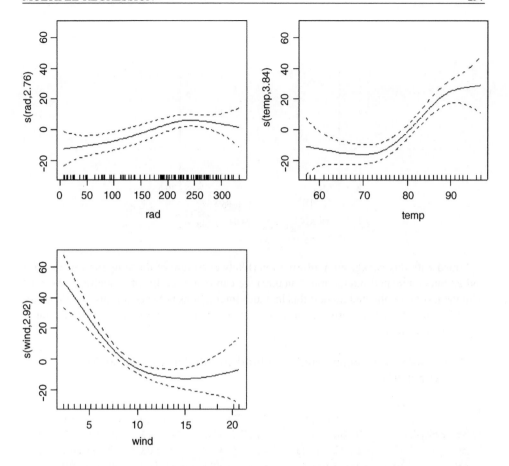

radiation. The next step might be to fit a tree model to see whether complex interactions between the explanatory variables are indicated:

```
library(tree)
model < -tree(ozone ~ .,data = ozone.pollution)
plot(model)
text(model)
```

This shows that temperature is far and away the most important factor affecting ozone concentration (the longer the branches in the tree, the greater the deviance explained). Wind speed is important at both high and low temperatures, with still air being associated with higher mean ozone levels (the figures at the ends of the branches are mean ozone concentrations). Radiation shows an interesting, but subtle effect. At low temperatures, radiation matters at relatively high wind speeds (>7.15), whereas at high temperatures, radiation matters at relatively low wind speeds (<10.6); in both cases, however, higher radiation is associated with higher mean ozone concentration. The tree model therefore indicates that the interaction structure of the data is not particularly complex (a reassuring finding).

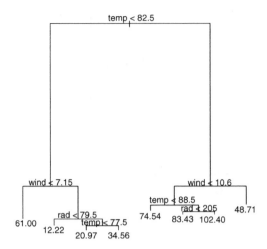

Armed with this background information (likely curvature of the temperature response and an uncomplicated interaction structure) we can begin the linear modelling. We start with the most complicated model: this includes interactions between all three explanatory variables plus quadratic terms to test for curvature in response to each of the three explanatory variables:

model1 < -lm(ozone ~ temp*wind*rad + I(rad^2) + I(temp^2) + I(wind^2))
summary(model1)

```
Coefficients:
                 Estimate    Std. Error   t value   Pr(>|t|)
(Intercept)      5.683e+02    2.073e+02    2.741    0.00725    **
temp            -1.076e+01    4.303e+00   -2.501    0.01401    *
wind            -3.237e+01    1.173e+01   -2.760    0.00687    **
rad             -3.117e-01    5.585e-01   -0.558    0.57799
I(rad^2)        -3.619e-04    2.573e-04   -1.407    0.16265
I(temp^2)        5.833e-02    2.396e-02    2.435    0.01668    *
I(wind^2)        6.106e-01    1.469e-01    4.157    6.81e-05   ***
temp:wind        2.377e-01    1.367e-01    1.739    0.08519.
temp:rad         8.402e-03    7.512e-03    1.119    0.26602
wind:rad         2.054e-02    4.892e-02    0.420    0.67552
temp:wind:rad   -4.324e-04    6.595e-04   -0.656    0.51358
```

```
Residual standard error: 17.82 on 100 degrees of freedom
Multiple R-Squared: 0.7394,      Adjusted R-squared: 0.7133
F-statistic: 28.37 on 10 and 100 DF, p-value:          0
```

The three-way interaction is clearly not significant, so we remove it to begin the process of model simplification:

model2 < -update(model1, ~. – temp:wind:rad)
summary(model2)

Next, we remove the least significant two-way interaction term – in this case wind:rad

```
model3 < -update(model2, ~ . – wind:rad)
summary(model3)
```

then try removing the temperature by wind interaction:

```
model4 < -update(model3, ~ . – temp:wind)
summary(model4)
```

We shall retain the marginally significant interaction between temp and rad ($p = 0.04578$) but leave out all other interactions. In model 4, the least significant quadratic term is for rad, so we delete this:

```
model5 < -update(model4, ~ . – I(rad^2))
summary(model5)
```

This deletion has rendered the temp:rad interaction insignificant, and caused the main effect of radiation to become insignificant. We should try removing the temp:rad interaction

```
model6 < -update(model5, ~ . – temp:rad)
summary(model6)
```

```
Coefficients:
                Estimate    Std. Error   t value   Pr(>|t|)
(Intercept)     291.16758   100.87723    2.886     0.00473    **
temp             -6.33955     2.71627   -2.334     0.02150     *
wind            -13.39674     2.29623   -5.834     6.05e-08   ***
rad               0.06586     0.02005    3.285     0.00139    **
I(temp^2)         0.05102     0.01774    2.876     0.00488    **
I(wind^2)         0.46464     0.10060    4.619     1.10e-05   ***
```

```
Residual standard error: 18.25 on 105 degrees of freedom
Multiple R-Squared: 0.713,          Adjusted R-squared: 0.6994
F-statistic: 52.18 on 5 and 105 DF, p-value:              0
```

Now we are making progress. All the terms in model 6 are significant. At this stage, we should check the assumptions, using plot(model6):

There is a clear pattern of variance increasing with the mean of the fitted values. This is bad news (heteroscedasticity). Also, the normality plot is distinctly curved; again, this is bad news. Let's try transformation of the response variable. There are no zeros in the response, so a log transformation is worth trying:

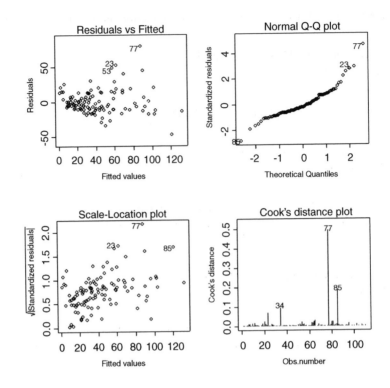

```
model7 < -lm(log(ozone) ~ temp + wind + rad + I(temp^2) + I(wind^2))
summary(model7)
```

```
Coefficients:
                Estimate    Std. Error   t value   Pr(>|t|)
(Intercept)     2.5538486    2.7359735     0.933    0.35274
temp           -0.0041416    0.0736703    -0.056    0.95528
wind           -0.2087025    0.0622778    -3.351    0.00112    **
rad             0.0025617    0.0005437     4.711    7.58e-06   ***
I(temp^2)       0.0003313    0.0004811     0.689    0.49255
I(wind^2)       0.0067378    0.0027284     2.469    0.01514    *
```

```
Residual standard error: 0.4949 on 105 degrees of freedom
Multiple R-Squared: 0.6882,      Adjusted R-squared: 0.6734
F-statistic: 46.36 on 5 and 105 DF, p-value:          0
```

On the log(ozone) scale, there is no evidence for a quadratic term in temperature, so let's remove that:

```
model8 < -update(model7, ~. − I(temp^2))
summary(model8)
```

```
Coefficients:
                Estimate    Std. Error   t value   Pr(>|t|)
(Intercept)     0.7231644   0.6457316    1.120     0.26528
temp            0.0464240   0.0059918    7.748     5.94e-12   ***
wind           -0.2203843   0.0597744   -3.687     0.00036    ***
rad             0.0025295   0.0005404    4.681     8.49e-06   ***
I(wind^2)       0.0072233   0.0026292    2.747     0.00706    **
```

Residual standard error: 0.4936 on 106 degrees of freedom
Multiple R-Squared: 0.6868,   Adjusted R-squared: 0.675
F-statistic: 58.11 on 4 and 106 DF, p-value: 0

plot(model8)

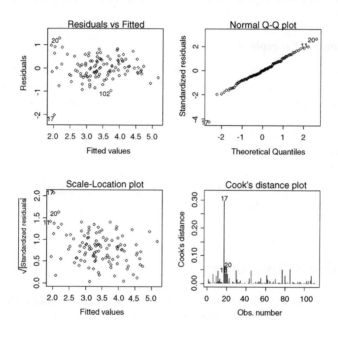

The heteroscedasticity and the non-normality have been cured, but there is now a highly influential data point (number 17 on the Cook's plot). We should refit the model with this point left out, to see if the parameter estimates or their standard errors are greatly affected:

```
model9 < -lm(log(ozone) ~ temp + wind + rad + I(wind^2),subset = (1:length
(ozone)! = 17))
summary(model9)
```

```
Coefficients:
                   Estimate    Std. Error   t value   Pr(>|t|)
(Intercept)        1.1932358   0.5990022    1.992     0.048963    *
temp               0.0419157   0.0055635    7.534     1.81e-11    ***
wind              -0.2208189   0.0546589   -4.040     0.000102    ***
rad                0.0022097   0.0004989    4.429     2.33e-05    ***
I(wind^2)          0.0068982   0.0024052    2.868     0.004993    **
Residual standard error: 0.4514 on 105 degrees of freedom
Multiple R-Squared: 0.6974,       Adjusted R-squared: 0.6859
F-statistic: 60.5 on 4 and 105 DF, p-value:                  0
```

Finally, plot(model9) shows that the variance and normality are well behaved, so we can stop at this point. We have found the minimal adequate model. It is on a scale of log(ozone concentration), all the main effects are significant, but there are no interactions, and there is a single quadratic term for wind speed (five parameters in all, with 105 d.f. for error).

A More Complex Example

In the next example we introduce two new difficulties: more explanatory variables and fewer data points. It is another air pollution dataframe, but the response variable in this case is sulphur dioxide concentration. There are six continuous explanatory variables:

```
pollute <-read.table("c:\\temp\\sulphur.dioxide.txt",header = T)
attach(pollute)
names(pollute)
```

```
[ 1] "Pollution"     "Temp"  "Industry"  "Population"  "Wind"
[ 6] "Rain"          "Wet.days"
```

Here are the 36 scatter plots:

```
pairs(pollute,panel = panel.smooth)
```

This time, let's begin with the tree model rather than the generalized additive model. A look at the pairs plots suggests that interactions may be more important than non-linearity in this case.

```
library(tree)
model <-tree(Pollution ~ .,data = pollute)
plot(model)
text(model)
```

This is interpreted as follows. The most important explanatory variable is Industry, and the threshold value separating low and high values of industry is 748. The right-hand branch of the tree indicates the mean value of air pollution for high levels of industry (67.00). The fact that this limb is unbranched means that no other variables explain a

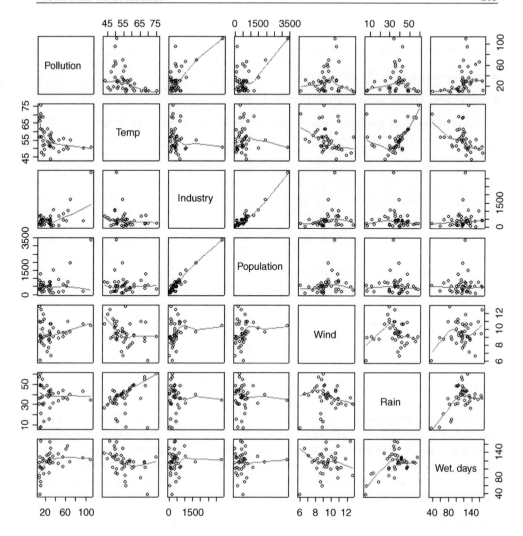

significant amount of the variation in pollution levels for high values of industry. The left-hand limb does not show the mean values of pollution for low values of industry, because there are other significant explanatory variables. Mean values of pollution are only shown at the extreme ends of branches. For low values of industry, the tree shows us that population has a significant impact on air pollution. At low values of population (<190) the mean level of air pollution was 43.43. For high values of population, the number of wet days is significant. Low numbers of wet days (<108) have mean pollution levels of 12.00 while temperature has a significant impact on pollution for places where the number of wet days is large. At high temperatures (>59.35 °F) the mean pollution level was 15.00 while at lower temperatures the run of wind is important. For still air (wind < 9.65) pollution was higher (33.88) than for higher wind speeds (23.00). The virtues of tree-based models are numerous:

- they are easy to appreciate and to describe to other people,

- the most important variables stand out,

- interactions are clearly displayed,

- non-linear effects are captured effectively, and

- the complexity of the behaviour of the explanatory variables is plain to see.

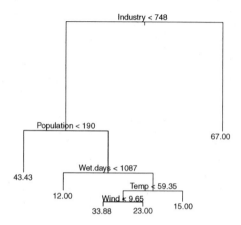

We conclude that the interaction structure is highly complex. We shall need to carry out the linear modelling with considerable care.

Start with some elementary calculations. With six explanatory variables, how many interactions might we fit? Well, there are $5 + 4 + 3 + 2 + 1 = 15$ two-way interactions for a start. Plus 20 three-way, 15 four-way and six five-way interactions, plus one six-way interaction for good luck. Then there are quadratic terms for each of the six explanatory variables. So we are looking at about 70 parameters that might be estimated from the data, but how many data points have we got?

length(Pollution)

[1] 41

Oh dear. We are planning to estimate almost twice as many parameters as there are data points. That's taking over-parameterization to new heights. We already know that you cannot estimate more parameter values than there are data points (i.e. a maximum of 41 parameters); but we also know that when we fit a saturated model to the data, it has no explanatory power (there are no degrees of freedom, so the model, by explaining everything, ends up explaining nothing at all). There is a useful rule of thumb: **don't try to estimate more than n/3 parameters during a multiple regression**. In the present case $n = 41$ so the rule of thumb is suggesting that we restrict ourselves to estimating about $41/3 \approx 13$ parameters at any one time. We know from the tree model that the interaction structure is going to be complicated so we shall concentrate on that. We begin, therefore, by looking for curvature, to see if we can eliminate it as a major cause of variation:

```
model1 <-
lm(Pollution ~ Temp + I(Temp^2) + Industry + I(Industry^2) + Population +
    I(Population^2) + Wind + I(Wind^2) + Rain + I(Rain^2) + Wet.days + I(Wet.days^2))
summary(model1)
```

```
Coefficients:
                     Estimate   Std. Error  t value   Pr(>|t|)
(Intercept)         -6.641e+01   2.234e+02   -0.297    0.76844
Temp                 5.814e-01   6.295e+00    0.092    0.92708
I(Temp^2)           -1.297e-02   5.188e-02   -0.250    0.80445
Industry             8.123e-02   2.868e-02    2.832    0.00847    **
I(Industry^2)       -1.969e-05   1.899e-05   -1.037    0.30862
Population          -7.844e-02   3.573e-02   -2.195    0.03662    *
I(Population^2)      2.551e-05   2.158e-05    1.182    0.24714
Wind                 3.172e+01   2.067e+01    1.535    0.13606
I(Wind^2)           -1.784e+00   1.078e+00   -1.655    0.10912
Rain                 1.155e+00   1.636e+00    0.706    0.48575
I(Rain^2)           -9.714e-03   2.538e-02   -0.383    0.70476
Wet.days            -1.048e+00   1.049e+00   -0.999    0.32615
I(Wet.days^2)        4.555e-03   3.996e-03    1.140    0.26398

Residual standard error: 14.98 on 28 degrees of freedom
Multiple R-Squared: 0.7148,        Adjusted R-squared: 0.5925
F-statistic: 5.848 on 12 and 28 DF,  p-value: 5.868e-005
```

So that's our first bit of good news. There is no evidence of curvature for any of the six explanatory variables. Only the main effects of industry and population are significant in this (over-parameterized) model. Now we need to consider the interaction terms. We do not fit interaction terms without both the component main effects, so we cannot fit all the two-way interaction terms at the same time (that would be $15 + 6 = 21$ parameters; well above the rule of thumb value of 13). One approach is to fit the interaction terms in randomly selected sets. With all six main effects, we can afford to assess $13 - 6 = 7$ interaction terms at a time, so we'll try this. Make a vector containing the names of the 15 two-way interactions:

```
interactions <- c("ti","tp","tw","tr","td","ip","iw","ir","id","pw","pr","pd","wr","wd", "rd")
```

Now shuffle the interactions into random order using **sample** without replacement:

```
sample(interactions)
```

```
[1] "wr" "wd" "id" "ir" "rd" "pr" "tp" "pw" "ti" "iw" "tw" "pd" "tr" "td" "ip"
```

It would be pragmatic to test the two-way interactions in three models each containing five two-way interaction terms:

model2 <-
lm(Pollution ~ Temp + Industry + Population + Wind + Rain + Wet.days + Wind:Rain +
    Wind: Wet.days + Industry:Wet.days + Industry:Rain + Rain:Wet.days)
model3 <-
lm(Pollution ~ Temp + Industry + Population + Wind + Rain + Wet.days + Population:
    Rain + Temp:Population + Population:Wind + Temp:Industry + Industry:Wind)
model4 <-
lm(Pollution ~ Temp + Industry + Population + Wind + Rain + Wet.days + Temp:Wind +
    Population:Wet.days + Temp:Rain + Temp:Wet.days + Industry:Population)

Extracting only the interaction terms from the three models, we see:

```
Industry:Rain          -1.616e-04    9.207e-04   -0.176   0.861891
Industry:Wet.days       2.311e-04    3.680e-04    0.628   0.534949
Wind:Rain               9.049e-01    2.383e-01    3.798   0.000690   ***
Wind:Wet.days          -1.662e-01    5.991e-02   -2.774   0.009593   **
Rain:Wet.days           1.814e-02    1.293e-02    1.403   0.171318

Temp:Industry          -1.643e-04    3.208e-03   -0.051   0.9595
Temp:Population         1.125e-03    2.382e-03    0.472   0.6402
Industry:Wind           2.668e-02    1.697e-02    1.572   0.1267
Population:Wind        -2.753e-02    1.333e-02   -2.066   0.0479     *
Population:Rain         6.898e-04    1.063e-03    0.649   0.5214

Temp:Wind               1.261e-01    2.848e-01    0.443   0.66117
Temp:Rain              -7.819e-02    4.126e-02   -1.895   0.06811.
Temp:Wet.days           1.934e-02    2.522e-02    0.767   0.44949
Industry:Population     1.441e-06    4.178e-06    0.345   0.73277
Population:Wet.days     1.979e-05    4.674e-04    0.042   0.96652
```

The next step might be to put all of the significant or close-to-significant interactions into the same model, and see which survive:

model5 <-
lm(Pollution ~ Temp + Industry + Population + Wind + Rain + Wet.days + Wind:Rain +
    Wind: Wet.days + Population:Wind + Temp:Rain)
summary(model5)

```
Coefficients:
                      Estimate    Std. Error   t value   Pr(>|t|)
(Intercept)         323.054546    151.458618     2.133   0.041226    *
Temp                 -2.792238      1.481312    -1.885   0.069153    .
Industry              0.073744      0.013646     5.404   7.44e-06  ***
Population            0.008314      0.056406     0.147   0.883810
Wind                -19.447031      8.670820    -2.243   0.032450    *
Rain                 -9.162020      3.381100    -2.710   0.011022    *
Wet.days              1.290201      0.561599     2.297   0.028750    *
Temp:Rain             0.017644      0.027311     0.646   0.523171
Population:Wind      -0.005684      0.005845    -0.972   0.338660
Wind:Rain             0.997374      0.258447     3.859   0.000562  ***
Wind:Wet.days        -0.140606      0.053582    -2.624   0.013530    *
```

We certainly do not need Temp:Rain

model6 <-update(model5, ~. –Temp:Rain)

or Population:Wind

model7 <-update(model6, ~. –Population:Wind)

All the terms in model 7 are significant. Time for a check on the behaviour of the model:

plot(model7)

That's not bad at all, but what about the higher-order interactions? One way to proceed is to specify the interaction level using ^3 in the model formula, but if you do this, you will find that we run out of degrees of freedom straight away. A sensible option is to fit three-way terms for the variables that already appear in two-way interactions – in our case, that is just one term: Wind:Rain:Wet.days

model8 <-update(model7, ~. + Wind:Rain:Wet.days)
summary(model8)

```
Coefficients:
                     Estimate    Std. Error   t value   Pr(>|t|)
(Intercept)         278.464474    68.041497     4.093   0.000282  ***
Temp                 -2.710981     0.618472    -4.383   0.000125  ***
Industry              0.064988     0.012264     5.299   9.1e-06   ***
Population           -0.039430     0.011976    -3.293   0.002485   **
Wind                 -7.519344     8.151943    -0.922   0.363444
Rain                 -6.760530     1.792173    -3.772   0.000685  ***
Wet.days              1.266742     0.517850     2.446   0.020311    *
Wind:Rain             0.631457     0.243866     2.589   0.014516    *
Wind:Wet.days        -0.230452     0.069843    -3.300   0.002440   **
Wind:Rain:Wet.days    0.002497     0.001214     2.056   0.048247    *
```

Residual standard error: 11.2 on 31 degrees of freedom
Multiple R-Squared: 0.8236,   Adjusted R-squared: 0.7724
F-statistic: 16.09 on 9 and 31 DF, p-value: 2.231e-009

That's enough for now. You are probably getting the idea. Multiple regression is difficult, time consuming and always vulnerable to subjective decisions about what to include and what to leave out. The linear modelling confirms the early impression from the tree model: for low levels of industry, the SO_2 level depends in a simple way on population (people tend to want to live where the air is clean) and in a complicated way on daily weather (the three-way interaction between wind, total rainfall and the number of wet days (i.e. on rainfall intensity).

Automating the Process of Model Simplification Using **step**

In a model with many interaction terms or a large number of explanatory variables, the procedure of model simplification can be very time-consuming. Help is at hand, however, in the form of the step function. The complex model1 is automatically simplified to model2 like this:

model2<-step(model1)

You can control whether the procedure steps "backwards", "forwards" or "both", and you can fix the most complex ("upper") and most simple ("lower") models between which simplification is carried out. The criterion used for dropping terms from the model is AIC; the smaller the AIC, the better the fit (see below).

Typically, step is generous in the sense that it leaves close-to-significant terms in the model. Therefore, the simplified model2 needs to be subjected to manual model simplification using update in order to arrive at a minimal adequate model which contains nothing but significant terms. You should not use step to simplify complex contingency table models without specifying "lower", because step could eliminate nuisance variables that need to be retained in the model to constrain the marginal totals.

AIC (Akaike's Information Criterion)

As you add parameters to a model you inevitably improve the fit. In the limit, you would have a parameter for every data point, and the fit of the model to the data would be perfect (see p. 153). There is always a trade-off between model simplicity and fit, and the ideal model is typically a compromise between these two. One way of determining whether extra parameters are justified is to use AIC. In the jargon, this is a penalized log-likelihood. It is like a deviance (−2*log-likelihood), but with a 'penalty' of 2 added to the score for every extra parameter in the model:

$$AIC = -2*\text{log-likelihood} + 2p$$

where p represents the number of parameters in the fitted model. It is useful in model simplification because a model with lower AIC is preferred to one with a higher AIC, and there are built-in tests for assessing the significance of the difference between two AICs. Unless an additional parameter causes a reduction in deviance of at least 2.0 then AIC will not decrease, and the additional parameter will not we warranted.

12

Contrasts

Contrasts are the essence of hypothesis testing and model simplification in Anova. They are used to compare means or groups of means with other means or groups of means, in what are known as **single degree of freedom comparisons**. There are two sorts of contrasts we might want to carry out:

- contrasts we had planned to carry out at the experimental design stage (these are referred to as *a priori* contrasts), or

- contrasts that look interesting after we have seen the results (these are referred to as *a posteriori* contrasts).

Some people are very snooty about *a posteriori* contrasts, on the grounds that they were **unplanned**. You are not supposed to decide what comparisons to make **after** you have seen the analysis, but scientists do this all the time – you cannot change human nature. The key point is that you should only do contrasts **after** the Anova has established that there really are significant differences to be investigated. It is not good practice to carry out tests to compare the largest mean with the smallest mean, if the Anova fails to reject the null hypothesis (tempting though this may be).

There are two important points to understand about contrasts:

- there is a huge number of **possible** contrasts,

- there are only $k - 1$ **orthogonal** contrasts.

where k is the number of factor levels. Two contrasts are said to be orthogonal to one another if the comparisons are statistically independent. Technically, two contrasts are orthogonal if **the products of their contrast coefficients sum to zero** (we shall see what this means in a moment).

Let's take a simple example. Suppose we have one factor with five levels and the factor levels are called a, b, c, d and e. Let's start writing down the possible contrasts. Obviously we could compare each mean singly with every other:

$$a \, vs. \, b, \; a \, vs. \, c, \; a \, vs. \, d, \; a \, vs. \, e, \; b \, vs. \, c, \; b \, vs. \, d, \; b \, vs. \, e, \; c \, vs. \, d, \; c \, vs. \, e, \; d \, vs. \, e$$

Statistics: An Introduction using R   M. J. Crawley
© 2005 John Wiley & Sons, Ltd   ISBNs: 0-470-02298-1 (PBK); 0-470-02297-3 (PPC)

but we could also compare pairs of means:

$\{a, b\}$ vs. $\{c, d\}$, $\{a, b\}$ vs. $\{c, e\}$, $\{a, b\}$ vs. $\{d, e\}$, $\{a, c\}$ vs. $\{b, d\}$, $\{a, c\}$ vs. $\{b, e\}$, etc.

or triplets of means:

$\{a, b, c\}$ vs. d, $\{a, b, c\}$ vs. e, $\{a, b, d\}$ vs. c, $\{a, b, d\}$ vs. e, $\{a, c, d\}$ vs. b, etc.

or groups of four means:

$\{a, b, c, d\}$ vs. e, $\{a, b, c, e\}$ vs. d, $\{b, c, d, e\}$ vs. a, $\{a, b, d, e\}$ vs. c, $\{a, b, c, e\}$ vs. d

You are probably getting the idea. There are absolutely masses of possible contrasts. In practice, however, we should only compare things once, either directly or implicitly. So the two contrasts:

$$a \text{ vs. } b \text{ and } a \text{ vs. } c$$

implicitly contrasts b vs. c. This means that if we have carried out the two contrasts a vs. b and a vs. c then the third contrast b vs. c is **not** an orthogonal contrast because you have already carried it out, implicitly. Which particular contrasts are orthogonal depends very much on your choice of the first contrast to make. Suppose there were good reasons for comparing $\{a,b,c,e\}$ vs. d. For example, d might be the placebo and the other four might be different kinds of drug treatment, so we make this our first contrast. Because $k - 1 = 4$ we only have three possible contrasts that are orthogonal to this. There may be *a priori* reasons to group $\{a,b\}$ and $\{c,e\}$ so we make this our second orthogonal contrast. This means that we have no degrees of freedom in choosing the last two orthogonal contrasts: they have to be a vs. b and c vs. e. Just remember that **with orthogonal contrasts you only compare things once**.

Contrast Coefficients

Contrast coefficients are a numerical way of embodying the hypothesis we want to test. The rules for constructing contrast coefficients are straightforward:

- treatments to be lumped together get the same sign (plus or minus),
- groups of means contrasted get the opposite sign,
- factor levels to be excluded get a contrast coefficient of 0,
- the contrast coefficients, c, must add up to 0.

Suppose that with our five-level factor $\{a,b,c,d,e\}$ we want to begin by comparing the four levels $\{a,b,c,e\}$ with the single level d. All levels enter the contrast, so none of the coefficients is 0. The four terms $\{a,b,c,e\}$ are grouped together so they all get the same sign (minus, for example, although it makes no difference which sign is chosen).

They are to be compared with d, so it gets the opposite sign (plus, in this case). The choice of what numeric values to give the contrast coefficients is entirely up to you. Most people use whole numbers rather than fractions, but it really doesn't matter. All that matters is that the c's add up to 0. The positive and negative coefficients have to add up to the same value. In our example, comparing four means with one mean, a natural choice of coefficients would be -1 for each of $\{a,b,c,e\}$ and $+4$ for d. Alternatively we could have selected $+0.25$ for each of $\{a,b,c,e\}$ and -1 for d.

factor level:	a	b	c	d	e
contrast one coefficients, c:	-1	-1	-1	4	-1

Suppose the second contrast is to compare $\{a,b\}$ with $\{c,e\}$. Because this contrast excludes d, we set its contrast coefficient to 0. $\{a,b\}$ get the same sign (say, plus) and $\{c,e\}$ get the opposite sign. Because the number of levels on each side of the contrast is equal (two in both cases) we can use the name numeric value for all the coefficients. The value 1 is the most obvious choice (but you could use 13.7 if you wanted to be perverse).

factor level:	a	b	c	d	e
contrast two coefficients, c:	1	1	-1	0	-1

There are only two possibilities for the remaining orthogonal contrasts: a vs. b and c vs. e:

factor level:	a	b	c	d	e
contrast three coefficients, c:	1	-1	0	0	0
contrast four coefficients, c:	0	0	1	0	-1

An Example of Contrasts in R

The example comes from the competition experiment we analysed in Chapter 9 in which the biomass of control plants is compared with the biomass of plants grown in conditions where competition was reduced in one of four different ways. There are two treatments in which the roots of neighbouring plants were cut (to 5 cm depth or 10 cm) and two treatments in which the shoots of neighbouring plants were clipped (25% or 50% of the neighbours cut back to ground level; see p. 167).

```
comp < -read.table("c:\\temp\\competition.txt",header = T)
attach(comp)
names(comp)

[ 1] "biomass"  "clipping"
```

We start with the one-way analysis of variance:

```
model1 < -aov(biomass ~ clipping)
summary(model1)
```

	Df	Sum Sq	Mean Sq	F value	Pr (>F)	
clipping	4	85356	21339	4.3015	0.008752	**
Residuals	25	124020	4961			

Clipping treatment has a highly significant effect on biomass – but have we fully understood the result of this experiment? Probably not. For example, which factor levels had the biggest effect on biomass, and were all of the competition treatments significantly different from the controls? To answer these questions, we need to use summary.lm:

summary.lm(model1)

Coefficients:

	Estimate	Std. Error	t value	Pr (>|t|)	
(Intercept)	465.17	28.75	16.177	9.33e-15	***
clippingn25	88.17	40.66	2.168	0.03987	*
clippingn50	104.17	40.66	2.562	0.01683	*
clippingr10	145.50	40.66	3.578	0.00145	**
clippingr5	145.33	40.66	3.574	0.00147	**

Residual standard error: 70.43 on 25 degrees of freedom
Multiple R-Squared: 0.4077,   Adjusted R-squared: 0.3129
F-statistic: 4.302 on 4 and 25 DF,   p-value: 0.008752

This looks as if we need to keep all five parameters, because all five rows of the summary table have one or more significance stars. In fact, this is not the case. This example highlights the major shortcoming of **treatment contrasts**: they do not show how many significant factor levels we need to retain in the minimal adequate model.

A Priori Contrasts

In this experiment, there are several planned comparisons we should like to make. The obvious place to start is by comparing the control plants that were exposed to the full rigours of competition, with all of the other treatments.

levels(clipping)

[1] "control" "n25" "n50" "r10" "r5"

That is to say, we want to contrast the first level of clipping with the other four levels. The contrast coefficients, therefore, would be 4, −1, −1, −1, −1. The next planned comparison might contrast the shoot-pruned treatments (n25 and n50) with the root-pruned treatments (r10 and r5). Suitable contrast coefficients for this would be 0, 1, 1, −1, −1 (because we are ignoring the control in this contrast). A third contrast might compare the two depths of root-pruning; 0, 0, 0, 1, −1. The last orthogonal contrast would therefore have to compare the two intensities of shoot-pruning: 0, 1, −1, 0, 0. Because the factor called 'clipping' has five levels there are only $5 - 1 = 4$ orthogonal contrasts.

R is outstandingly good at dealing with contrasts, and we can associate these five user-specified *a priori* contrasts with the categorical variable called clipping like this:

```
contrasts(clipping) <-
cbind(c(4,–1,–1,–1,–1),c(0,1,1,–1,–1),c(0,0,0,1, –1),c(0,1, –1,0,0))
```

We can check that this has done what we wanted by typing

```
contrasts(clipping)
```

	[,1]	[,2]	[,3]	[,4]
control	4	0	0	0
n25	-1	1	0	1
n50	-1	1	0	-1
r10	-1	-1	1	0
r5	-1	-1	-1	0

which produces the matrix of contrast coefficients that we specified. Note that all the columns add to zero (i.e. each set of contrast coefficients is correctly specified). Note also that the products of any two of the columns sum to zero (this shows that all the contrasts are orthogonal, as intended), e.g. comparing contrasts 1 and 2 gives products $0 + (-1) + (-1) + 1 + 1 = 0$.

Now we can re-fit the model and inspect the results of our specified contrasts, rather than the default treatment contrasts:

```
model2 < -aov(biomass ~ clipping)
summary.lm(model2)
```

```
Coefficients:
             Estimate  Std. Error  t value    Pr(>|t|)
(Intercept)  561.80000   12.85926   43.688     <2e-16   ***
clipping1    -24.15833    6.42963   -3.757    0.000921  ***
clipping2    -24.62500   14.37708   -1.713    0.099128  .
clipping3      0.08333   20.33227    0.004    0.996762
clipping4     -8.00000   20.33227   -0.393    0.697313

Residual standard error: 70.43 on 25 degrees of freedom
Multiple R-Squared: 0.4077,      Adjusted R-squared: 0.3129
F-statistic: 4.302 on 4 and 25 DF,   p-value: 0.008752
```

Instead of requiring five parameters (as suggested by out initial treatment contrasts), this analysis shows that we need only two parameters: the overall mean (561.8) and the contrast between the controls and the four competition treatments ($p = 0.000921$). All the other contrasts are non-significant.

Model Simplification by Step-wise Deletion

An alternative to specifying the contrasts ourselves (as above) is to aggregate non-significant factor levels in a step-wise *a posteriori* procedure. To demonstrate this, we revert to treatment contrasts:

```
contrasts(clipping) < -NULL
options(contrasts = c("contr.treatment","contr.poly"))
```

Now we fit the model with all five factor levels as a starting point:

```
model3 < -aov(biomass ~ clipping)
summary.lm(model3)
```

```
Coefficients:
                Estimate  Std. Error  t value   Pr (>|t|)
(Intercept)      465.17      28.75    16.177    9.33e-15  ***
clippingn25       88.17      40.66     2.168    0.03987   *
clippingn50      104.17      40.66     2.562    0.01683   *
clippingr10      145.50      40.66     3.578    0.00145   **
clippingr5       145.33      40.66     3.574    0.00147   **
```

Looking down the list of parameter estimates, we see that the most similar are the effects of root pruning to 10 and 5 cm (145.5 vs. 145.33). We shall begin by simplifying these to a single root-pruning treatment called root. The trick is to use 'levels gets' to change the names of the appropriate factor levels. Start by copying the original factor name:

```
clip2 < -clipping
```

Now inspect the level numbers of the various factor level names:

```
levels(clip2)
[ 1] "control"  "n25"  "n50"  "r10"  "r5"
```

The plan is to lump together r10 and r5 under the same name, 'root'. These are the fourth and fifth levels of clip2, so we write:

```
levels(clip2)[4:5] < -"root"
```

and to see what has happened type

```
levels(clip2)
[ 1] "control"  "n25"  "n50"  "root"
```

and we see that 'r10' and 'r5' have indeed been replaced by 'root'. The next step is to fit a new model with clip2 in place of clipping, and to test whether the new simpler model is significantly worse as a description of the data using anova:

```
model4 <-aov(biomass ~ clip2)
anova(model3,model4)
```

```
Analysis of Variance Table
Model 1: biomass~clipping
Model 2: biomass~clip2
Res.Df     RSS   Df    Sum of Sq              F      Pr (>F)
1    25  124020
2    26  124020   -1   -0.0833333   0.0000168      0.9968
```

As we expected, this model simplification was completely justified. The next step is to investigate the effects using summary.lm:

```
summary.lm(model4)
```

```
Coefficients:
                Estimate  Std. Error  t value   Pr (>|t|)
(Intercept)       465.17       28.20   16.498   2.66e-15  ***
clip2n25           88.17       39.87    2.211   0.036029    *
clip2n50          104.17       39.87    2.612   0.014744    *
clip2root         145.42       34.53    4.211   0.000269  ***
```

It looks as if the two shoot clipping treatments (n25 and n50) are not significantly different from one another (they differ by just 16.0 with a standard error of 39.87). We can lump these together into a single shoot-pruning treatment as follows:

```
clip3 <-clip2
levels(clip3)[2:3] <-"shoot"
levels(clip3)
```

```
[ 1] "control"  "shoot"  "root"
```

Then fit a new model with clip3 in place of clip2:

```
model5 <-aov(biomass ~ clip3)
anova(model4,model5)
```

```
Analysis of Variance Table

Model 1: biomass~clip2
Model 2: biomass~clip3
     Res. Df      RSS   Df   Sum of Sq       F  Pr (>F)
1         26   124020
2         27   124788   -1        -768   0.161  0.6915
```

Again, this simplification was fully justified. Do the root and shoot competition treatments differ?

```
clip4 < -clip3
levels(clip4)[2:3] < -"pruned"
levels(clip4)
```

```
[ 1]  "control"  "pruned"
```

Now fit a new model with clip4 in place of clip3:

```
model6 < -aov(biomass ~ clip4)
anova(model5,model6)
```

```
Analysis of Variance Table
```

```
Model 1: biomass~clip3
Model 2: biomass~clip4
      Res.Df      RSS   Df  Sum of Sq        F    Pr (>F)
1          27   124788
2          28   139342   -1     -14553   3.1489   0.08726.
```

This simplification was close to significant, but we are ruthless ($p > 0.05$, so we accept the simplification). Now we have the minimal adequate model:

```
summary.lm(model6)
```

```
Coefficients:
                Estimate  Std. Error  t value   Pr (>|t|)
(Intercept)        465.2        28.8   16.152   1.11e-15  ***
clip4pruned        120.8        32.2    3.751   0.000815  ***
```

it has just two parameters: the mean for the controls (465.2) and the difference between the control mean and the four treatment means ($465.2 + 120.8 = 586.0$):

```
tapply(biomass,clip4,mean)
```

```
  control      pruned
465.1667   585.9583
```

We know that these two means are significantly different from the p value $= 0.000815$, but just to show how it is done, we can make a final model 7 that has no explanatory variable at all (it fits only the overall mean). This is achieved by writing $y \sim 1$ in the model formula:

```
model7 < -aov(biomass ~ 1)
anova(model6,model7)
```

```
Analysis of Variance Table
```

```
Model 1: biomass~clip4
Model 2: biomass~1
```

	Res.Df	RSS	Df	Sum of Sq	F	Pr (>F)
1	28	139342				
2	29	209377	-1	-70035	14.073	0.000815 ★★★

Note that the p value is exactly the same as in model 6. The p values in R are calculated such that they avoid the need for this final step in model simplification: they are 'p on deletion' values.

Contrast Sums of Squares by Hand

The key point to understand is that **the treatment sum of squares SSA is the sum of all $(k-1)$ orthogonal sums of squares**. It is useful to know which of the contrasts contributes most to SSA, and to work this out, we compute the contrast sum of squares SSC as follows:

$$SSC = \frac{\left(\sum \frac{c_i T_i}{n_i}\right)^2}{\sum \frac{c_i^2}{n_i}}.$$

The significance of a contrast is judged in the usual way by carrying out an F test to compare the contrast variance with the error variance, s^2. Since all contrasts have a single degree of freedom, the contrast variance is equal to SSC, so the F test is just

$$F = \frac{SSC}{s^2},$$

where the error variance, s^2, comes from the error mean square column of the Anova table. The contrast is significant (i.e. the two contrasted groups have significantly different means) if the calculated value is larger than the critical value of F with one and $k(n-1)$ degrees of freedom. We demonstrate these ideas by continuing our example.
   The five mean biomass values were:

tapply(biomass,clipping,mean)

control	n25	n50	r10	r5
465.1667	553.3333	569.3333	610.6667	610.5

We have already established that the contrast between the controls and the other four treatments was highly significant (above). Here we develop the theme by assessing the significance of the type of competition treatment. The root pruned plants (r10 and r5) were larger than the shoot pruned plants (n25 and n50), suggesting that below ground competition might be more influential than above ground. It remains to be seen whether these differences are significant by using contrasts. To compare defoliation and root pruning (i.e. a comparison of competition for light with below-ground competition), the contrast coefficients are

	control	n25	n50	r10	r5
c_i	0	-1	-1	1	1

To calculate a new contrast sum of squares, we need the treatment totals, T,

tapply(biomass,clipping,sum)

```
control    n25     n50     r10      r5
   2791   3320    3416    3664    3663
```

to which we apply the formula. The controls have zero weight so we ignore them.

$$SSC = \frac{\left[\frac{1}{6}(-1 \times 3320) + (-1 \times 3416) + (1 \times 3664) + (1 \times 3663)\right]^2}{\frac{1}{6}((-1)^2 + (-1)^2 + 1^2 + 1^2)} = \frac{\left(\frac{591}{6}\right)^2}{\frac{4}{6}} = 14553.38.$$

The error variance is 4960.81 (from the Anova table, above), so the F test for this contrast is

$$F = \frac{14553.38}{4960.81} = 2.93367.$$

Notice that this F value is the square of the t value obtained by contrast number 2, above ($1.7128^2 = 2.933684$). We need to test the significance of this by comparing our calculated F value with the critical value with 1 and 25 d.f.. We use qf for this

qf(0.95,1,25)

[1] 4.241699

Our calculated value is less than the value in tables, so this contrast was not significant.

Comparison of the Three Kinds of Contrasts

In order to show the differences between treatment, Helmert and sum contrasts, we shall reanalyse this competition experiment.

1. Treatment contrasts

This is the default in R. These are the contrasts you get, unless you explicitly choose otherwise.

options(contrasts = c("contr.treatment","contr.poly"))

Here are the contrast coefficients as set under treatment contrasts

contrasts(clipping)

	n25	n50	r10	r5
control	0	0	0	0
n25	1	0	0	0

n50	0	1	0	0
r10	0	0	1	0
r5	0	0	0	1

Notice that the contrasts are **not** orthogonal (the products of the coefficients do not sum to zero).

output.treatment < -lm(biomass ~ clipping)
summary(output.treatment)

Coefficients:

	Estimate	Std. Error	t value	Pr (>|t|)	
(Intercept)	465.17	28.75	16.177	9.33e-15	***
clippingn25	88.17	40.66	2.168	0.03987	*
clippingn50	104.17	40.66	2.562	0.01683	*
clippingr10	145.50	40.66	3.578	0.00145	**
clippingr5	145.33	40.66	3.574	0.00147	**

With treatment contrasts, the factor levels are arranged in alphabetical sequence, and the level that comes first in the alphabet is made into the intercept. In our example this is 'control', so we can read off the control mean as 465.17, and the standard error of a mean as 28.75. The remaining four rows are differences between means, and the standard errors are standard errors of differences. Thus, clipping neighbours back to 25 cm increases biomass by 88.17 over the controls and this difference is significant at $p = 0.03987$. And so on. The downside of treatment contrasts is that all the rows appear to be significant despite the fact that rows 2–5 are actually not significantly different from one another, as we saw earlier.

2. Helmert contrasts

This is the default in S-Plus, so beware if you are switching back and forth between the two languages.

options(contrasts = c("contr.helmert","contr.poly"))

contrasts(clipping)

	[, 1]	[, 2]	[, 3]	[, 4]
control	-1	-1	-1	-1
n25	1	-1	-1	-1
n50	0	2	-1	-1
r10	0	0	3	-1
r5	0	0	0	4

Notice that the contrasts are orthogonal (the products sum to zero) and their coefficients sum to zero, unlike treatment contrasts, above.

output.helmert < -lm(biomass ~ clipping)
summary(output.helmert)

```
Coefficients:
               Estimate    Std. Error    t value    Pr (> |t|)
(Intercept) 561.800          12.859      43.688     <2e-16       ***
clipping1    44.083          20.332       2.168     0.0399         *
clipping2    20.028          11.739       1.706     0.1004
clipping3    20.347           8.301       2.451     0.0216         *
clipping4    12.175           6.430       1.894     0.0699         .
```

With Helmert contrasts, the intercept is the overall mean (561.8). The first contrast (on row 2, labelled contrast '1') compares the first mean with the average of the first and second factor levels in alphabetical sequence (control plus n25, see above); its parameter value is the mean of the first two factor levels, minus the mean of the first factor level:

(465.16667 + 553.33333)/2-465.166667

[1] 44.08332

The third row contains the contrast between the third factor level (n50) and the two levels already compared (control and n25); its value is the difference between the average of the first three factor levels and the average of the first two factor levels

(465.16667 + 553.33333 + 569.333333)/3-(465.166667 + 553.3333)/2

[1] 20.02779

The fourth row contains the contrast between the fourth factor level (r10) and the three levels already compared (control, n25 and n50); its value is the difference between the average of the first four factor levels and the average of the first three factor levels

(465.16667 + 553.33333 + 569.333333 + 610.66667)/4-
(553.3333 + 465.166667 + 569.3333)/3

[1] 20.34725

The fifth and final row contains the contrast between the fifth factor level (r5) and the four levels already compared (control, n25, n50 and r10); its value is the difference between the average of the first five factor levels, and the average of the first four factor levels

mean(biomass)-(465.16667 + 553.33333 + 569.333333 + 610.66667)/4

[1] 12.175

So much for the parameter estimates. Now look at the standard errors. We have seen none of these values in any of the analyses we have done to date. The standard error in row 1 is the standard error of the overall mean, with s^2 taken from the overall Anova table: $\sqrt{\frac{s^2}{k.n}}$

sqrt(4961/30)

[1] 12.85950

The standard error in row 2 is a comparison of **a group of two means with a single mean** $(2 \times 1 = 2)$. This is multiplied by the sample size n in the denominator: $\sqrt{\frac{s^2}{2 \times n}}$

sqrt(4961/(2*6))

[1] 20.33265

The standard error in row 3 is a comparison of **a group of three means with a group of two means** $(3 \times 2 = 6)$: $\sqrt{\frac{s^2}{6 \times n}}$

sqrt(4961/(3*2*6))

[1] 11.73906

The standard error in row 4 is a comparison of **a group of four means with a group of three means** $(4 \times 3 = 12)$: $\sqrt{\frac{s^2}{12 \times n}}$

sqrt(4961/(4*3*6))

[1] 8.30077

The standard error in row 5 is a comparison of **a group of five means with a group of four means** $(5 \times 4 = 20)$: $\sqrt{\frac{s^2}{20 \times n}}$

sqrt(4961/(5*4*6))

[1] 6.429749

It is true that the parameter estimates and their standard errors are much more difficult to understand in Helmert than in treatment contrasts. However, the advantage of Helmert contrasts is that they give you proper orthogonal contrasts, and hence give a much clearer picture of which factor levels need to be retained in the minimal adequate model. They do not eliminate the need for careful model simplification, however. As we saw earlier, this example requires only two parameters in the minimal adequate model, but Helmert contrasts (above) suggest the need for three (albeit only marginally significant) parameters.

3. Sum contrasts

options(contrasts = c("contr.sum","contr.poly"))

I do not know anyone who uses sum contrasts, so I won't use up space explaining them. If you are interested, see Statistical Computing, Crawley 2002, p. 341.

Aliasing

Aliasing occurs when there is no information available on which to base an estimate of a parameter value. Parameters can be aliased for one of two reasons:

- there are no data in the dataframe from which to estimate the parameter (e.g. missing values, partial designs or correlation amongst the explanatory variables), or

- the model is structured in such a way that the parameter value cannot be estimated (e.g. over specified models with more parameters than necessary).

Intrinsic aliasing occurs when it is due to **the structure of the model**. **Extrinsic aliasing** occurs when it is due to **the nature of the data**.

If we had a factor with four levels (say none, light, medium and heavy use) then we could estimate four means from the data, one for each factor level. But the model looks like this:

$$y = \mu + \beta_1 x_1 + \beta_2 x_2 + \beta_3 x_3 + \beta_4 x_4$$

where the x's are dummy variables having the value 0 or 1 (see p. 164). Clearly there is no point in having five parameters in the model if we can estimate only four independent terms. One of the parameters must be intrinsically aliased.

There are innumerable ways of dealing with this, but three equally logical options are:

- set the grand mean μ to 0, so that the four β's are the four individual treatment means,

- set the first term β_1 to 0 so that μ is the mean of the first group and the β's are the differences between the first group mean and the other group means,

- set the sum of the β's to 0 so that μ is the grand mean and each β is a departure from the grand mean.

Suppose that in a factorial experiment, all of the animals receiving level 2 of diet (factor A) and level 3 of temperature (factor B) have died accidentally as a result of attack by a fungal pathogen. This particular combination of diet and temperature contributes no data to the response variable, so the interaction term A(2):B(3) cannot be estimated. It is **extrinsically aliased**, and its parameter estimate is set to zero. If one continuous variable is perfectly correlated with another variable that has already been fitted to the data (perhaps because it is a constant multiple of the first variable), then the second term is aliased and adds nothing to the model. Suppose that $x_2 = 0.5x_1$ then fitting a model with $x_1 + x_2$ will lead to x_2 being **intrinsically aliased** and given a zero parameter estimate (see the example of galls on leaves, above). If all the values of a particular explanatory variable are set to zero for a given level of a particular factor, then that level is **intentionally aliased**. This sort of aliasing is a useful programming trick in Ancova when we wish a covariate to be fitted to some levels of a factor but not to others.

Contrasts and the Parameters of Ancova Models

In analysis of covariance, we estimate a slope and an intercept for each level of one or more factors. Suppose we are modelling growth (the response variable) as a function of gender and age. Gender is a factor with two levels (male and female) and age is a continuous measure. The maximal model therefore has four parameters: two slopes (a slope for males and a slope for females) and two intercepts (one for males and one for females) like this:

$$weight_{male} = a_{male} + b_{male} \times age$$
$$weight_{female} = a_{female} + b_{female} \times age$$

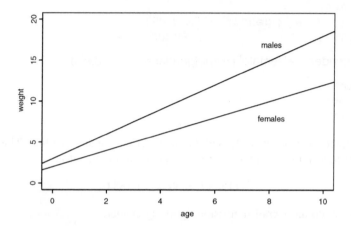

The difficulty arises because there are several different ways of expressing the values of the four parameters in the summary.lm table:

- two slopes, and two intercepts (as in the equations, above),

- one slope and one difference between slopes, and one intercept and one difference between intercepts, or

- the overall mean slope and the overall mean intercept, and one difference between slopes and one difference between intercepts.

In the second case (two estimates and two differences) a decision needs to be made about which factor level to associate with the estimate, and which level with the difference (e.g. should males be expressed as the intercept and females as the difference between intercepts, or vice versa)? When the factor levels are unordered (the typical case), then R takes the factor level that comes first in the alphabet as the estimate and the others are expressed as differences. In our example, the parameter estimates would be female, and male parameters would be expressed as differences from the female values, because 'f' comes before 'm' in the alphabet. This should become clear from an example:

```
Ancovacontrasts<-read.table("c:\\temp\\Ancovacontrasts.txt",header=T)
attach(Ancovacontrasts)
names(Ancovacontrasts)
```

```
[ 1]  "weight"   "gender"   "age"
```

First we work out the two regressions separately so that we know the values of the two slopes and the two intercepts:

```
lm(weight[gender=="male"]~age[gender=="male"])
```

```
Coefficients:
  Intercept)   age[ gender == "male"]
    3.115178                1.560808
```

```
lm(weight[gender=="female"]~age[gender= ="female"])
```

```
Coefficients:
  (Intercept)     age[ gender == "female"]
    1.966277                 0.9962039
```

So the intercept for males is 3.115 and the intercept for females is 1.966. The difference between the first (female) and second intercepts (male) is therefore

$$3.115 - 1.9266 = +1.1884.$$

Now we can do an overall regression, ignoring gender:

```
lm(weight~age)
```

```
Coefficients:
  (Intercept)            age
    2.540728      1.278506
```

This tells us that the average intercept is 2.541 and the average slope is 1.279.

Next we can carry out an analysis of covariance and compare the output produced by each of the three different contrast options allowed by S-Plus: **Helmert** (the default), **treatment** (the default in R and in Glim) and **sum**.

```
options(contrasts=c("contr.helmert", "contr.poly"))
```

The Ancova estimates separate slopes and intercepts for each gender because we use the asterisk operator:

```
lm(weight~age*gender)
```

```
Coefficients:
  (Intercept)           age           gender    age:gender
    2.540728      1.278506      0.5744508     0.2823018
```

Let's see if we can work out what the four parameter values represent. The first parameter 2.5407 (labelled 'Intercept') is the intercept of the overall regression, ignoring gender (see above). The parameter labelled age (1.2785) is a **slope** because age is our continuous explanatory variable. Again, you will see that it is the slope for the regression of weight against age, ignoring gender. The third parameter labelled **gender** (0.5744) must have something to do with intercepts because **gender** is our categorical variable. If we want to reconstruct the second intercept (for males) we need to add 0.5744 to the overall intercept: $2.5407 + 0.5744 = 3.1151$. To get the intercept for females we need to subtract it $2.5407 - 0.5744 = 1.9663$. The fourth parameter (0.2823) labelled **age:gender** is the difference between the overall mean slope (1.279) and the male slope: $1.2785 + 0.2823 = 1.5608$. To get the slope of weight against age for females we need to subtract the interaction term from the age term: $1.2785 - 0.2823 = 0.9962$.

The advantage of Helmert contrasts is in hypothesis testing, because it is easy to see which terms we need to retain in a simplified model by inspecting their significance levels in the summary.lm table. The disadvantage is that it is hard to reconstruct the slopes and the intercepts from the estimated parameters values (see also p. 220). Let's repeat the analysis using **treatment contrasts** as used by R and by Glim:

```
options(contrasts = c("contr.treatment", "contr.poly"))

lm(weight ~ age*gender)

Coefficients:
  (Intercept)         age       gender   age:gender
     1.966277   0.9962039     1.148902    0.5646037
```

The Intercept (1.9662) is now the intercept for females (because 'f' comes before 'm' in the alphabet). The **age** parameter (0.9962) is the slope of the graph of weight against age for females. The **gender** parameter (1.1489) is the difference between the (female) intercept and the male intercept $(1.966277 + 1.148902 = 3.1151)$. The **age:gender** interaction term is the difference between slopes of the female and male graphs $(0.9962 + 0.5646 = 1.5608)$. So with treatment contrasts, the parameters (in order 1 to 4) are an intercept, a slope, a difference between two intercepts, and a difference between two slopes. Many people are more comfortable with this method of presentation than they are with Helmert contrasts.

Finally, we look at the third option which is **sum contrasts**:

```
options(contrasts = c("contr.sum", "contr.poly"))

lm(weight ~ age*gender)

Coefficients:
  (Intercept)        age       gender    age:gender
     2.540728   1.278506   -0.5744508    -0.2823018
```

The first two terms are the same as those produced by Helmert contrasts: the overall intercept and slope of the graph relating weight to age ignoring gender. The gender parameter (-0.5744) is **sign reversed** compared with the Helmert option: it shows how to calculate the female (the **first**) intercept from the overall intercept $2.5407 - 0.5746 = 1.9661$. The interaction term also has reversed sign – to get the slope for females, add the interaction term to the slope for age: $1.2785 - 0.2823 = 0.9962$.

Multiple Comparisons

The thorny issue of multiple comparisons arises because when we do more than one test we are likely to find 'false positives' at an inflated rate (i.e. by rejecting a true null hypothesis more often than α). The old-fashioned approach was to use Bonferroni's correction; in looking up a value for Student's t, you divide your α value by twice the number of comparisons you have done. If the result is still significant then all is well, but it often will not be. Bonferroni's correction is very harsh and will often throw out the baby with the bathwater. An old-fashioned alternative was to use Duncan's Multiple Range Tests (you may have seen these in old statistics books, where lower case letters were written at the head of each bar in a barplot: bars with different letters were significantly different, while bars with the same letter were not significantly different). The modern approach is to use contrasts wherever possible, and where it is essential to do multiple comparisons, then to use the wonderfully named Tukey's Honest Significant Differences (see Statistical Computing, Crawley 2002), and see

?TukeyHSD

13

Count Data

Up to this point, the response variables have all been continuous measurements like weights, heights, lengths, temperatures, growth rates and so on. A great deal of the data collected by scientists, medical statisticians and economists, however, are in the form of *counts* (whole numbers or integers). The number of individuals that died, the number of firms going bankrupt, the number of days of frost, the number of red blood cells on a microscope slide, or the number of craters in a sector of lunar landscape are all potentially interesting variables for study. With count data, the number 0 often appears as a value of the response variable (consider, for example, what a 0 would mean in the context of the examples just listed). In this chapter we deal with data on **frequencies**, where we count how many times something happened, but we have no way of knowing how often it did **not** happen (e.g. lightening strikes, bankruptcies, deaths, births). This is in contrast with count data on **proportions**, where we know the number doing a particular thing, but also the number not doing that thing (e.g. the proportion dying, gender ratios at birth, proportions of different groups responding to a questionnaire).

Straightforward linear regression methods (assuming constant variance, normal errors) are not appropriate for count data for four main reasons:

- the linear model might lead to the prediction of negative counts,
- the variance of the response variable is likely to increase with the mean,
- the errors will not be normally distributed, and
- zeros are difficult to handle in transformations.

In R, count data are handled very elegantly in a glm by specifying family = poisson which sets errors = Poisson and link = log (see Chapter 7). The log link ensures that all the fitted values are positive, while the Poisson errors take account of the fact that the data are integers and have variances that are equal to their means.

A Regression with Poisson Errors

This example has a count (the number of reported cancer cases per year per clinic) as the response variable, and a single continuous explanatory variable (the distance from a

Statistics: An Introduction using R   M. J. Crawley
© 2005 John Wiley & Sons, Ltd   ISBNs: 0-470-02298-1 (PBK); 0-470-02297-3 (PPC)

nuclear plant to the clinic in km). The question is whether or not proximity to the reactor affects the number of cancer cases.

```
clusters <-read.table("c:\\temp\\clusters.txt",header = T)
attach(clusters)
names(clusters)
```

```
[ 1] "Cancers"  "Distance"
```

```
plot(Distance,Cancers)
```

There seems to be a downward trend in cancer cases with distance (see the plot below), but is the trend significant? We do a regression of cases against distance, using a glm with Poisson errors:

```
model1 <-glm(Cancers ~ Distance,poisson)
summary(model1)
```

```
Coefficients:
                Estimate   Std. Error   z value  Pr(>|z|)
(Intercept)     0.186865    0.188728     0.990    0.3221
Distance       -0.006138    0.003667    -1.674    0.0941 .
```

```
(Dispersion parameter for poisson family taken to be 1)
```

```
    Null deviance: 149.48 on 93 degrees of freedom
Residual deviance: 146.64 on 92 degrees of freedom
AIC: 262.41
```

The trend does not look to be significant, but first look at the residual deviance. It is assumed that this is the same as the residual degrees of freedom. The fact that residual deviance is larger than residual degrees of freedom indicates that we have overdispersion (extra, unexplained variation in the response). We compensate for the overdispersion by re-fitting the model using quasipoisson rather than Poisson errors:

```
model2 <-glm(Cancers ~ Distance,quasipoisson)
summary(model2)
```

```
Coefficients:
                Estimate   Std. Error   t value  Pr(>|t|)
(Intercept)     0.186865    0.235341     0.794    0.429
Distance       -0.006138    0.004573    -1.342    0.183
```

```
(Dispersion parameter for quasipoisson family taken to be
1.554966)
```

```
    Null deviance: 149.48 on 93 degrees of freedom
Residual deviance: 146.64 on 92 degrees of freedom
AIC: NA
```

Compensating for the overdispersion has increased the p value to 0.183, so there is no compelling evidence to support the existence of a trend in cancer incidence with distance from the nuclear plant. To draw the fitted model through the data, you need to understand that the glm with Poisson errors uses the log link, so the parameter estimates and the predictions from the model (the 'linear predictor') are in logs, and need to be antilogged before the (non-significant) fitted line is drawn.

```
xv <-seq(0,100,.1)
yv <-predict(model2,list(Distance = xv))
lines(xv,exp(yv))
```

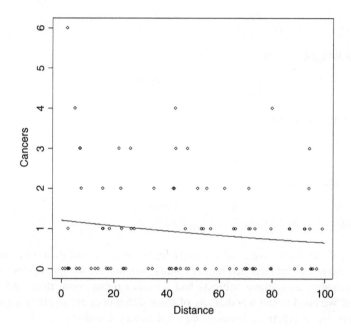

Analysis of Deviance with Count Data

The response variable is a count of infected blood cells per mm^2 on microscope slides prepared from randomly selected individuals. The explanatory variables are smoker (logical, yes or no), age (three levels, under 20, 21 to 59, 60 and over), gender (male or female) and body mass score (three levels: normal, overweight, obese).

```
count <-read.table("c:\\temp\\cells.txt",header = T)
attach(count)
names(count)

[ 1] "cells"    "smoker"    "age"    "gender"    "weight"
```

It is always a good idea with count data to get a feel for the overall frequency distribution of counts using table:

table(cells)

0	1	2	3	4	5	6	7
314	75	50	32	18	13	7	2

Most subjects (314 of them) showed no damaged cells, and the maximum of seven was observed in just two patients. We begin data inspection by tabulating the main effect means:

tapply(cells,smoker,mean)

FALSE	TRUE
0.5478723	1.9111111

tapply(cells,weight,mean)

normal	obese	over
0.5833333	1.2814371	0.9357143

tapply(cells,gender,mean)

female	male
0.6584507	1.2202643

tapply(cells,age,mean)

mid	old	young
0.8676471	0.7835821	1.2710280

It looks as if smokers have a substantially higher mean count than non-smokers, that overweight and obese subjects had higher counts than normal weight, males had a higher count that females, and young subjects had a higher mean count than middle-aged or older people. We need to test whether any of these differences are significant and to assess whether there are interactions between the explanatory variables.

model1 < -glm(cells ~ smoker*gender*age*weight,poisson)
summary(model1)

```
     Null deviance:  1052.95 on 510 degrees of freedom
Residual deviance:   736.33 on 477 degrees of freedom
AIC: 1318
```

Number of Fisher Scoring iterations: 6

The residual deviance (736.33) is much greater than the residual degrees of freedom (477) indicating overdispersion, so before interpreting any of the effects, we should re-fit the model using quasipoisson errors:

model2 < -glm(cells ~ smoker*gender*age*weight,quasipoisson)
summary(model2)

```
Call:
glm(formula = cells~smoker * gender * age * weight, family = quasipoisson)

Deviance Residuals:
   Min        1Q      Median       3Q        Max
-2.236    -1.022     -0.851     0.520      3.760

Coefficients: (2 not defined because of singularities)
```

	Estimate Std.	Error	t value	Pr(>|t|)
(Intercept)	-0.8329	0.4307	-1.934	0.0537 .
smokerTRUE	-0.1787	0.8057	-0.222	0.8246
gendermale	0.1823	0.5831	0.313	0.7547
ageold	-0.1830	0.5233	-0.350	0.7267
ageyoung	0.1398	0.6712	0.208	0.8351
weightobese	1.2384	0.8965	1.381	0.1678
weightover	-0.5534	1.4284	-0.387	0.6986
smokerTRUE:gendermale	0.8293	0.9630	0.861	0.3896
smokerTRUE:ageold	-1.7227	2.4243	-0.711	0.4777
smokerTRUE:ageyoung	1.1232	1.0584	1.061	0.2892
gendermale:ageold	-0.2650	0.9445	-0.281	0.7791
gendermale:ageyoung	-0.2776	0.9879	-0.281	0.7788
smokerTRUE:weightobese	3.5689	1.9053	1.873	0.0617 .
smokerTRUE:weightover	2.2581	1.8524	1.219	0.2234
gendermale:weightobese	-1.1583	1.0493	-1.104	0.2702
gendermale:weightover	0.7985	1.5256	0.523	0.6009
ageold:weightobese	-0.9280	0.9687	-0.958	0.3386
ageyoung:weightobese	-1.2384	1.7098	-0.724	0.4693
ageold:weightover	1.0013	1.4776	0.678	0.4983
ageyoung:weightover	0.5534	1.7980	0.308	0.7584
smokerTRUE:gendermale:ageold	1.8342	2.1827	0.840	0.4011
smokerTRUE:gendermale:ageyoung	-0.8249	1.3558	-0.608	0.5432
smokerTRUE:gendermale:weightobese	-2.2379	1.7788	-1.258	0.2090
smokerTRUE:gendermale:weightover	-2.5033	2.1120	-1.185	0.2365
smokerTRUE:ageold:weightobese	0.8298	3.3269	0.249	0.8031
smokerTRUE:ageyoung:weightobese	-2.2108	1.0865	-2.035	0.0424 *
smokerTRUE:ageold:weightover	1.1275	1.6897	0.667	0.5049
smokerTRUE:ageyoung:weightover	-1.6156	2.2168	-0.729	0.4665
gendermale:ageold:weightobese	2.2210	1.3318	1.668	0.0960 .
gendermale:ageyoung:weightobese	2.5346	1.9488	1.301	0.1940
gendermale:ageold:weightover	-1.0641	1.9650	-0.542	0.5884
gendermale:ageyoung:weightover	-1.1087	2.1234	-0.522	0.6018
smokerTRUE:gendermale:ageold:weightobese	-1.6169	3.0561	-0.529	0.5970
smokerTRUE:gendermale:ageyoung:weightobese	NA	NA	NA	NA
smokerTRUE:gendermale:ageold:weightover	NA	NA	NA	NA
smokerTRUE:gendermale:ageyoung:weightover	2.4160	2.6846	0.900	0.3686

```
- - -
Signif. codes:  0 `***'  0.001 `**'  0.01 `*'  0.05 `.'  0.1 ` '  1

(Dispersion parameter for quasipoisson family taken to be 1.854809)

    Null deviance: 1052.95 on 510 degrees of freedom
Residual deviance:  736.33 on 477 degrees of freedom
```

There is an apparently significant three-way interaction between smoking, age and obesity ($p = 0.0424$). There were too few subjects to assess the four-way interaction (see the NAs in the table) so we begin model simplification by removing the highest-order interaction:

```
model3 < -update(model2, ~ . - smoker:gender:age:weight)
summary(model3)
```

```
Call:
glm(formula = cells~smoker + gender + age + weight + smoker:gender +
smoker:age + gender:age + smoker:weight + gender:weight + age:weight +
smoker:gender:age + smoker:gender:weight + smoker:age:weight +
gender:age:weight, family = quasipoisson)
```

```
Deviance Residuals:
    Min        1Q      Median        3Q        Max
 -2.2442   -1.0477    -0.8921     0.5195     3.7613
```

Coefficients:

	Estimate	Std. Error	t value	Pr(>|t|)	
(Intercept)	-0.897195	0.436987	-2.053	0.04060	*
smokerTRUE	0.030263	0.735384	0.041	0.96719	
gendermale	0.297192	0.570008	0.521	0.60234	
ageold	-0.118726	0.528164	-0.225	0.82224	
ageyoung	0.289259	0.639617	0.452	0.65130	
weightobese	1.302660	0.898306	1.450	0.14768	
weightover	-0.005052	1.027197	-0.005	0.99608	
smokerTRUE:gendermale	0.527345	0.867292	0.608	0.54345	
smokerTRUE:ageold	-0.566584	1.700587	-0.333	0.73915	
smokerTRUE:ageyoung	0.757297	0.939745	0.806	0.42073	
gendermale:ageold	-0.379884	0.935363	-0.406	0.68482	
gendermale:ageyoung	-0.610703	0.920967	-0.663	0.50758	
smokerTRUE:weightobese	3.924591	1.475474	2.660	0.00808	**
smokerTRUE:weightover	1.192159	1.259886	0.946	0.34450	
gendermale:weightobese	-1.273202	1.040700	-1.223	0.22178	
gendermale:weightover	0.154097	1.098779	0.140	0.88853	
ageold:weightobese	-0.993355	0.970483	-1.024	0.30656	
ageyoung:weightobese	-1.346913	1.459452	-0.923	0.35653	
ageold:weightover	0.454217	1.090258	0.417	0.67715	
ageyoung:weightover	-0.483955	1.300863	-0.372	0.71004	
smokerTRUE:gendermale:ageold	0.771116	1.451509	0.531	0.59549	
smokerTRUE:gendermale:ageyoung	-0.210317	1.140383	-0.184	0.85376	
smokerTRUE:gendermale:weightobese	-2.500668	1.369939	-1.825	0.06857	.
smokerTRUE:gendermale:weightover	-1.110222	1.217529	-0.912	0.36230	
smokerTRUE:ageold:weightobese	-0.882951	1.187869	-0.743	0.45766	
smokerTRUE:ageyoung:weightobese	-2.453315	1.047065	-2.343	0.01954	*
smokerTRUE:ageold:weightover	0.823018	1.528230	0.539	0.59045	
smokerTRUE:ageyoung:weightover	0.040795	1.223662	0.033	0.97342	
gendermale:ageold:weightobese	2.338617	1.324803	1.765	0.07816	.
gendermale:ageyoung:weightobese	2.822032	1.623846	1.738	0.08288	.
gendermale:ageold:weightover	-0.442066	1.545449	-0.286	0.77497	
gendermale:ageyoung:weightover	0.357807	1.291192	0.277	0.78181	

The remaining model simplification is left to you as an exercise. Your minimal adequate model might look something like this:

```
summary(model18)
```

```
Call:
glm(formula = cells~smoker + weight + smoker:weight, family =
quasipoisson)

Deviance Residuals:
    Min          1Q      Median        3Q        Max
-2.6511      -1.1742    -0.9148    0.5533    3.6436

Coefficients:
                         Estimate  Std. Error   t value   Pr(>|t|)
(Intercept)               -0.8712      0.1760    -4.950   1.01e-06  ***
smokerTRUE                 0.8224      0.2479     3.318   0.000973  ***
weightobese                0.4993      0.2260     2.209   0.027598    *
weightover                 0.2618      0.2522     1.038   0.299723
smokerTRUE:weightobese     0.8063      0.3105     2.597   0.009675   **
smokerTRUE:weightover      0.4935      0.3442     1.434   0.152225

(Dispersion parameter for quasipoisson family taken to be 1.827925)

    Null deviance:     1052.95 on 510 degrees of freedom
Residual deviance:      792.85 on 505 degrees of freedom
```

This model shows a highly significant interaction between smoking and weight in determining the number of damaged cells, but there are no convincing effects of age or gender. In a case like this, it is useful to produce a summary table to highlight the effects:

```
tapply(cells,list(smoker,weight),mean)
```

```
            normal         obese           over
FALSE    0.4184397     0.689394      0.5436893
TRUE     0.9523810     3.514286      2.0270270
```

The interaction arises because the response to smoking depends on body weight: smoking adds a mean of about 0.5 damaged cells for individuals with normal body weight, but adds 2.8 damaged cells for obese people.
   It is straightforward to turn the summary table into a barplot:

```
barplot(tapply(cells,list(smoker,weight),mean),beside=T)
legend(1.2,3.4,c("non","smoker"),fill=c(2,7))
```

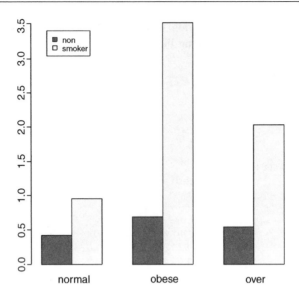

The Danger of Contingency Tables

We have already dealt with simple contingency tables and their analysis using Fisher's Exact Test or Pearson's chi-squared (see p. 90), but there is an important further issue to be dealt with. In observational studies we quantify only a limited number of explanatory variables. It is inevitable that we shall fail to measure a number of factors that have an important influence on the behaviour of the system in question. That's life, and given that we make every effort to note the important factors, there is little we can do about it. The problem comes when we ignore factors that have an important influence on the response variable. This difficulty can be particularly acute if we **aggregate data over important explanatory variables**. An example should make this clear.

Suppose we are carrying out a study of induced defences in trees. A preliminary trial has suggested that early feeding on a leaf by aphids may cause chemical changes in the leaf which reduce the probability of that leaf being attacked later in the season by hole-making insects. To this end we mark a large cohort of leaves, then score whether they were infested by aphids early in the season and whether they were holed by insects later in the year. The work was carried out on two different trees and the results were as follows:

Tree	Aphids	Holed	Intact	Total leaves	Proportion holed
Tree 1	Without	35	1750	1785	0.0196
	With	23	1146	1169	0.0197
Tree 2	Without	146	1642	1788	0.0817
	With	30	333	363	0.0826

There are four variables: the response variable, count, with eight values (in grey above), a two-level factor for late season feeding by caterpillars (holed or intact), a two-level factor for early season aphid feeding (with or without aphids) and a two-level factor for tree (the observations come from two separate trees, imaginatively named Tree 1 and Tree 2).

```
induced < -read.table("C:\\temp\\induced.txt",header = T)
attach(induced)
names(induced)
```

```
[ 1]  "Tree"     "Aphid"     "Caterpillar"     "Count"
```

We begin by fitting what is known as a **saturated model**. This is a curious thing, which has as many parameters as there are values of the response variable. The fit of the model is perfect, so there are no residual degrees of freedom and no residual deviance. The reason that we fit a saturated model is that it is always the best place to start modelling complex contingency tables. If we fit the saturated model, then there is no risk that we inadvertently leave out important interactions between the so-called 'nuisance variables'. These are the parameters that need to be in the model to ensure that the marginal totals are properly constrained.

```
model < -glm(Count ~ Tree*Aphid*Caterpillar,family = poisson)
```

The asterisk notation ensures that the saturated model is fitted, because all of the main effects and two-way interactions are fitted, along with the three-way interaction Tree by Aphid by Caterpillar. The model fit involves the estimation of $2 \times 2 \times 2 = 8$ parameters, and exactly matches the eight values of the response variable, Count. There is no point looking at the saturated model in any detail, because the reams of information it contains are all superfluous. The first real step in the modelling is to use update to remove the three-way interaction from the saturated model, and then to use anova to test whether the three-way interaction is significant or not.

```
model2 < -update(model, ~ . - Tree:Aphid:Caterpillar)
```

The punctuation here is very important (it is comma, tilde, dot, minus) and note the use of colons rather than asterisks to denote interaction terms rather than main effects plus interaction terms. Now we can see whether the three-way interaction was significant by specifying test = "Chi" like this:

```
anova(model,model2,test = "Chi")
```

```
Analysis of Deviance Table
```

```
Model 1: Count~Tree * Aphid * Caterpillar
Model 2: Count~Tree + Aphid + Caterpillar + Tree:Aphid +
  Tree:Caterpillar + Aphid:Caterpillar
```

Resid. Df	Resid. Dev	Df	Deviance	P(>|Chi|)
1 0	-9.97e-14			
2 1	0.00079	-1	-0.00079	0.97756

This shows clearly that the interaction between caterpillar attack and leaf holing does not differ from tree to tree ($p = 0.97756$). Note that if this interaction had been significant, then we would have stopped the modelling at this stage, but it wasn't so we leave it out and continue. What about the main question – is there an interaction between caterpillar attack and leaf holing? To test this we delete the Caterpillar:Aphid interaction from the model, and assess the results using anova:

```
model3 <-update(model2, ~. - Aphid:Caterpillar)
anova(model3,model2,test = "Chi")
Analysis of Deviance Table
Model 1: Count~Tree + Aphid + Caterpillar + Tree:Aphid +
   Tree:Caterpillar
Model 2: Count~Tree + Aphid + Caterpillar + Tree:Aphid +
   Tree:Caterpillar + Aphid:Caterpillar
```

Resid. Df	Resid. Dev	Df	Deviance	P(>|Chi|)
1	2	0.00409		
2	1	0.00079	1	0.00329	0.95423

There is absolutely no hint of an interaction ($p = 0.954$). The interpretation is clear: this work provides no evidence for induced defences caused by early season caterpillar feeding.

However, look what happens when we do the modelling the wrong way. Suppose we went straight for the interaction of interest, Aphid:Caterpillar. We might proceed like this:

```
wrong <-glm(Count ~ Aphid*Caterpillar,family = poisson)
wrong1 <-update(wrong, ~. - Aphid:Caterpillar)
anova(wrong,wrong1,test = "Chi")
Analysis of Deviance Table
Model 1: Count~Aphid * Caterpillar
Model 2: Count~Aphid + Caterpillar
```

Resid. Df	Resid. Dev	Df	Deviance	P(>|Chi|)
1	4	550.19		
2	5	556.85	-1	-6.66	0.01

The Aphid:Caterpillar interaction is highly significant ($p = 0.01$) providing strong evidence for induced defences. This is **wrong** ! By failing to include Tree in the model we have omitted an important explanatory variable. As it turns out, and we should really have determined by more thorough preliminary analysis, the trees differ enormously in their average levels of leaf holing:

```
as.vector(tapply(Count,list(Caterpillar,Tree),sum))[1]/tapply(Count,Tree,sum) [1]
Tree1
0.01963439
```

as.vector(tapply(Count,list(Caterpillar,Tree),sum))[3]/ tapply(Count,Tree,sum) [2]

```
Tree2
0.08182241
```

   Tree 2 has more than four times the proportion of its leaves holed by caterpillars. If we had been paying more attention when we did the modelling the wrong way, we should have noticed that the model containing only Aphid and Caterpillar had massive overdispersion, and this should have alerted us that all was not well. The moral is simple and clear. Always fit a saturated model first, containing all the variables of interest and all the interactions involving the nuisance variables (Tree in this case). Only delete from the model those interactions that involve the variables of interest (Aphid and Caterpillar in this case). Main effects are meaningless in contingency tables, as are the model summaries. Always test for overdispersion. It will never be a problem if you follow the advice of simplifying down from a saturated model, because you only ever leave out non-significant terms, and you never delete terms involving any of the nuisance variables.

Analysis of Covariance with Count Data

In this example the response is a count of the number of plant species on plots that have different biomass (a continuous explanatory variable) and different soil pH (a categorical variable with three levels: high, mid and low).

```
species < -read.table("c:\\temp\\species.txt",header = T)
attach(species)
names(species)
```

```
[ 1]   "pH"     "Biomass"     "Species"
```

```
plot(Biomass,Species,type = "n")
spp < -split(Species,pH)
bio < -split(Biomass,pH)
points(bio[[1]],spp[[1]],pch = 16)
points(bio[[2]],spp[[2]],pch = 17)
points(bio[[3]],spp[[3]])
```

Note the use of split to create separate lists of plotting coordinates for the three levels of pH. It is clear that species declines with biomass, and that soil pH has a big effect on species, but does the slope of the relationship between species and biomass depend on pH? The lines look reasonably parallel from the scatter plot. This is a question about interaction effects, and in analysis of covariance, interaction effects are about differences between slopes:

```
model1 < -glm(Species ~ Biomass*pH,poisson)
summary(model1)
```

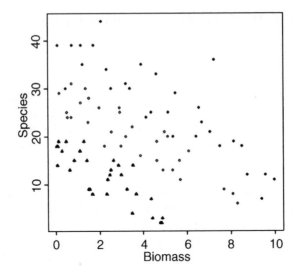

```
Coefficients:
                     Estimate   Std. Error   z value    Pr(>|z|)
(Intercept)           3.76812      0.06153    61.242     <2e-16 ***
Biomass              -0.10713      0.01249    -8.579     <2e-16 ***
pHlow                -0.81558      0.10282    -7.932    2.16e-15 ***
pHmid                -0.33146      0.09216    -3.596    0.000323 ***
Biomass:pHlow        -0.15502      0.03996    -3.880    0.000105 ***
Biomass:pHmid        -0.03189      0.02307    -1.382    0.166885
```

(Dispersion parameter for poisson family taken to be 1)

```
    Null deviance: 452.346 on 89 degrees of freedom
Residual deviance:  83.201 on 84 degrees of freedom
AIC:     514.39
```

We can test for the need for different slopes by comparing this maximal model (with six parameters) with a simpler model with different intercepts but the same slope (four parameters):

```
model2 <-glm(Species ~ Biomass + pH,poisson)
anova(model1,model2,test = "Chi")
```

Analysis of Deviance Table

```
Model 1: Species~Biomass * pH
Model 2: Species~Biomass + pH
```

	Resid. Df	Resid. Dev	Df	Deviance	P(>|Chi|)
1	84	83.201			
2	86	99.242	-2	-16.040	0.0003288

The slopes are very significantly different ($p = 0.00033$), so we are justified in retaining the more complicated model 1. Finally, we draw the fitted lines through the scatterplot, using predict:

```
xv <-seq(0,10,0.1)
levels(pH)
```

```
[ 1] "high"     "low"     "mid"
```

```
length(xv)
```

```
[ 1] 101
```

```
phv <-rep("high",101)
yv <-predict(model1,list(pH = factor(phv),Biomass = xv),type = "response")
lines(xv,yv)
phv <-rep("mid",101)
yv <-predict(model1,list(pH = factor(phv),Biomass = xv),type = "response")
lines(xv,yv)
phv <-rep("low",101)
yv <-predict(model1,list(pH = factor(phv),Biomass = xv),type = "response")
lines(xv,yv)
```

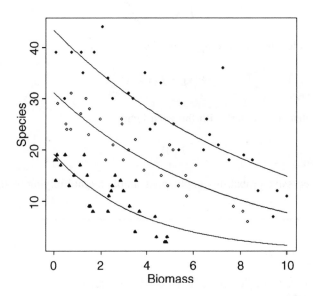

Note the use of type = "response" in the predict function. This ensures that yv is calculated as species rather than log(species), and means we do not need to back-transform using antilogs before drawing the lines (compare with the example on p. 229). You could make the R code more elegant by writing a function to plot any number of lines, depending on the number of levels of the factor (three levels of pH in this case).

Frequency Distributions

Here are data on the numbers of bankruptcies in 80 districts. The question is whether there is any evidence that some districts show greater than expected numbers of cases. What would we expect? Of course we should expect some variation, but how much, exactly? Well that depends on our model of the process. Perhaps the simplest model is that absolutely nothing is going on, and that every single bankruptcy case is absolutely independent of every other. That leads to the prediction that the numbers of cases per district will follow a Poisson process – a distribution in which the variance is equal to the mean (Box 13.1).

Box 13.1. The Poisson distribution

The Poisson distribution is widely used for the description of count data. We know how many times something happened (e.g. kicks from cavalry horses, lightening strikes, bomb hits), but we have no way of knowing how many times it did not happen. The Poisson is a one-parameter distribution, specified entirely by the mean. The variance is identical to the mean (λ), so the variance/mean ratio is equal to one. The probability of observing a count of x is given by

$$P(x) = \frac{e^{-\lambda}\lambda^x}{x!}.$$

This can be calculated very simply on a hand calculator because:

$$P(x) = P(x-1)\frac{\lambda}{x}.$$

This means that if you start with the **zero term**

$$P(0) = e^{-\lambda}$$

then each successive probability is obtained simply by multiplying by the mean and dividing by x.

Let's see what the data show.

```
case.book <-read.table("c:\\temp\\cases.txt",header=T)
attach(case.book)
names(case.book)

[1]  "cases"
```

First we need to count the numbers of districts with no cases, one case, two cases, and so on. The R function that does this is called table:

```
frequencies < -table(cases)
frequencies
```

```
cases
  0     1     2     3     4     5     6     7     8     9     10
 34    14    10     7     4     5     2     1     1     1      1
```

There were no cases at all in 34 districts, but one district had ten cases. A good way to proceed is to compare our distribution (called frequencies) with the distribution that would be observed if the data really did come from a Poisson distribution as postulated by our model. We can use the R function dpois to compute the probability density of each of the 11 frequencies from 0 to 10 (we multiply the probability produced by dpois by the total sample of 80 to obtain the predicted frequencies). We need to calculate the mean number of cases per district: this is the Poisson distribution's only parameter:

```
mean(cases)
```

```
[ 1]  1.775
```

The plan is to draw two distributions side by side, so we set up the plotting region:

```
par(mfrow = c(1,2))
```

Now we plot the observed frequencies in the left-hand panel:

```
barplot(frequencies,ylab = "Frequency",xlab = "Cases",col = "red")
```

and the predicted, Poisson frequencies in the right-hand panel

```
barplot(dpois(0:10,1.775)*80,names = as.character(0:10),ylab = "Frequency",
   xlab = "Cases",col = "red")
```

The distributions are very different (p. 242) the mode of the observed data is zero, but the mode of the Poisson with the same mean is one; the observed data contained examples of eight, nine and ten cases, but these would be highly unlikely under a Poisson process. We would say that the observed data are highly **aggregated**; they have a variance/mean ratio much greater than one (the Poisson, of course, has variance/mean ratio $= 1$):

```
var(cases)/mean(cases)
```

```
[ 1]  2.99483
```

So, if the data are not Poisson distributed, how are they distributed? A good candidate distribution where the variance/mean ratio is this big (about 3.0) is the negative binomial distribution (Box 13.2).

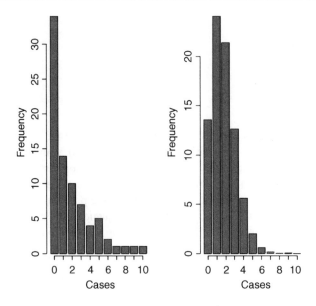

Box 13.2. Negative binomial distribution

This discrete, two-parameter distribution is useful for describing the distribution of count data, where the variance is often much greater than the mean. The two parameters are the mean μ and the clumping parameter k. The smaller the value of k, the greater the degree of clumping. The density function is

$$p(x) = \left(1 + \frac{\mu}{k}\right)^{-k} \frac{\Gamma(k+x)}{x!\Gamma(k)} \left(\frac{\mu}{\mu+k}\right)^x$$

where Γ is the gamma function. The zero term is found by setting $x = 0$ and simplifying:

$$p(0) = \left(1 + \frac{\mu}{k}\right)^{-k}$$

and successive terms in the distribution can be computed iteratively from

$$p(x) = p(x-1)\left(\frac{k+x-1}{x}\right)\left(\frac{\mu}{\mu+k}\right).$$

An initial estimate of the value of k can be obtained from the sample mean and variance

$$k \approx \frac{\mu^2}{s^2 - \mu}.$$

Since k cannot be negative, it is clear that the negative binomial distribution should not be fitted to data where the variance is less than the mean (use a binomial distribution instead). The precise maximum likelihood estimate of k is found numerically, by iterating progressively more fine-tuned values of k until the left- and right-hand sides of the following equation are equal:

$$n \ln\left(1 + \frac{\mu}{k}\right) = \sum_{x=0}^{max}\left(\frac{A_{(x)}}{k+x}\right)$$

where the vector $A(x)$ contains the total frequency of values **greater** than x. You could create a function to work out the probability densities like this:

```
factorial <-function(x) max(cumprod(1:x))
negbin <-function(x,u,k) (1+u/k)^(-k)*(u/(u+k))^x*gamma(k+x)/(factorial(x)
   *gamma(k))
```

then use the function to produce a barplot of probability densities for a range of x values (say 0 to 10, for a distribution with specified mean and aggregation parameter (say $\mu = 0.8, k = 0.2$) like this

```
xf <-numeric(11)
for (i in 0:10) xf[i+1] <-negbin(i,0.8,0.2)
barplot(xf)
```

This is a two-parameter distribution; the first parameter is the mean number of cases (1.775), and the second is called the clumping parameter, k (measuring the degree of aggregation in the data: small values of k ($k < 1$) show high aggregation, while large values of k ($k > 5$) show randomness). We can get an approximate estimate of the magnitude of k from

$$k = \frac{\mu^2}{s^2 - \mu}.$$

We can work this out:

```
mean(cases)^2/(var(cases)-mean(cases))
```

```
[1]  0.8898003
```

so we shall work with $k = 0.89$. How do we compute the expected frequencies? The density function for the negative binomial distribution is dnbinom and it has three arguments: the frequency for which we want the probability (in our case 0 to 10), the number of successes (in our case 1), and the mean number of cases (1.775); we multiply by the total number of cases (80) to obtain the expected frequencies

```
exp <-dnbinom(0:10,1,mu=1.775)*80
```

The plan is to draw a single figure in which the observed and expected frequencies are drawn side by side. The trick is to produce a new vector (called both) which is twice as long as the observed and expected frequency vectors ($2 \times 11 = 22$). Then, we put the observed frequencies in the odd numbered elements (using modulo 2 to calculate the values of the subscripts), and the expected frequencies in the even numbered elements:

```
both < -numeric(22)
both[1:22 %% 2 != 0] < -frequencies
both[1:22 %% 2 == 0] < -exp
```

On the *x* axis, we intend to label only every other bar:

```
labels < -character(22)
labels[1:22 %% 2 == 0] < -as.character(0:10)
```

Now we can produce the barplot, using white for the observed frequencies and grey for the negative binomial frequencies:

```
barplot(both,col = rep(c("white","grey"),11),names = labels,
    ylab = "Frequency", xlab = "Cases")
```

Now we need to add a legend to show what the two colours of the bars mean. You can locate the legend by trial and error, or by left-clicking mouse when the cursor is in the correct position, using the locator(1) function (see p. 126):

```
legend(16,30,c("Observed","Expected"), fill = c("white","grey"))
```

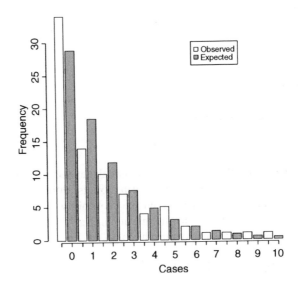

The fit to the negative binomial distribution is much better than it was with the Poisson distribution, especially in the right-hand tail; but the observed data have too many zeros and too few ones to be represented perfectly by a negative binomial distribution. If you want to quantify the lack of fit between the observed and expected frequency distributions, you can calculate Pearson's chi square $\sum (O - E)^2 / E$ based on the number of comparisons that have expected frequency > 4

exp

```
[ 1]    28.8288288    18.4400617    11.7949944    7.5445460    4.8257907    3.0867670
[ 7]     1.9744185     1.2629164     0.8078114    0.5167082    0.3305070
```

If we accumulate the rightmost six frequencies, then all the values of exp will be bigger than four. The degrees of freedom are then given by the number of comparisons (six) – the number of parameters estimated from the data (two in our case) – 1 (for contingency, because the total frequency must add up to 80) = three degrees of freedom. We use 'levels gets' to reduce the lengths of the observed and expected vectors, creating an upper interval called '5+' for '5 or more':

```
cs < -factor(0:10)
levels(cs)[6:11] < -"5 + "
levels(cs)
```

```
[ 1]    "0"      "1"      "2"      "3"      "4"      "5+"
```

Now make the two shorter vectors 'of' and 'ef' (for observed and expected frequencies):

```
ef < -as.vector(tapply(exp,cs,sum))
of < -as.vector(tapply(frequencies,cs,sum))
```

Finally we can compute the chi square value measuring the difference between the observed and expected frequency distributions, and use 1-pchisq to work out the *p* value:

```
sum((of-ef)^2/ef)
```

```
[ 1]    3.594145
```

```
1-pchisq(3.594145,3)
```

```
[ 1]    0.3087555
```

We conclude that a negative binomial description of these data is reasonable (the observed and expected distributions are not significantly different; $p = 0.31$).

14

Proportion Data

An important class of problems involves data on proportions such as:

- studies on percentage mortality,

- infection rates of diseases,

- proportion responding to clinical treatment,

- proportion admitting to particular voting intentions,

- sex ratios, or

- data on proportional response to an experimental treatment.

What all these have in common is that we know how many of the experimental objects are in one category (say dead, insolvent, male or infected) and we also know how many are in another (say alive, solvent, female or uninfected). This contrasts with Poisson count data, where we knew how many times an event occurred, but **not** how many times it did not occur (Chapter 13).

We model processes involving proportional response variables in R by specifying a glm with family = binomial (Box 14.1). The only complication is that whereas with Poisson errors we could simply say family = poisson, with binomial errors we must specify the number of failures as well as the numbers of successes in a two-vector response variable. To do this we bind together two vectors using cbind into a single object, y, comprising the numbers of successes and the number of failures. The *binomial denominator, n,* is the total sample, and the

number.of.failures = binomial.denominator − number.of.successes

y <- cbind(number.of.successes, number.of.failures)

The old-fashioned way of modelling these sort of data was to use the percentage mortality as the response variable. There are four problems with this:

Statistics: An Introduction using R   M. J. Crawley
© 2005 John Wiley & Sons, Ltd   ISBNs: 0-470-02298-1 (PBK); 0-470-02297-3 (PPC)

- the errors are not normally distributed,

- the variance is not constant,

- the response is bounded (by 1 above and by 0 below), and

- by calculating the percentage, we lose information of the size of the sample, n, from which the proportion was estimated.

Box 14.1 The binomial distribution

This is a 1-parameter distribution in which the parameter p describes the probability of success in a Bernoulli trial with outcomes 1 or 0. The probability of x successes out of n attempts is given by multiplying together

- the probability of obtaining one specific realisation $p^x(1-p)^{n-x}$

- the number of ways of getting that realisation

The number of ways of getting x items out of n items is given by the combinatorial formula

$$\binom{n}{x} = \text{ways of getting } x \text{ out of } n = \frac{n!}{x!(n-x)!}$$

where ! means 'factorial'. For instance, $5! = 5 \times 4 \times 3 \times 2 = 120$. The R function for this is called choose(n,x). The density function of the binomial distribution is

$$p(x) = \binom{n}{x}p^x(1-p)^{n-x}$$

evaluated by the function dbinom in R. The mean of the binomial distribution is np and the variance is $np(1-p)$. Since $(1-p)$ is less than 1 it is obvious that *the variance is less than the mean* for the binomial distribution (it is useful for describing regular patterns; cf. the negative binomial (p. 242) which is useful for describing aggregated patterns).

R carries out weighted regression, using the individual sample sizes as weights, and the logit link function to ensure linearity. There are some kinds of proportion data, like **percentage cover**, which are best analysed using conventional models (normal errors and constant variance) following **arc-sine transformation**. The response variable, y, measured in radians, is $\sin^{-1}\sqrt{0.01 \times p}$ where p is percentage cover. If, however, the response variable takes the form of a **percentage change** in some continuous measurement (such as the percentage change in weight on receiving a particular diet), then rather than arc-sine transform the data, it is usually better treated by:

- either analysis of covariance (see Chapter 9), using final weight as the response variable and initial weight as a covariate, or

- by specifying the response variable as a relative growth rate, measured as log(final weight/initial weight),

both of which can be analysed with normal errors without further transformation.

Analyses of Data on One and Two Proportions

For comparisons of one binomial proportion with a constant, use binom.test (see p. 83). For comparison of two samples of proportion data, use prop.test (see p. 84). The methods of this chapter are required only for more complex models of proportion data, including regression and contingency tables, where generalized linear models are used.

Count Data on Proportions

The traditional transformations of proportion data were arcsine and probit. The arcsine transformation took care of the error distribution, while the probit transformation was used to linearize the relationship between percentage mortality and log dose in a bioassay. There is nothing wrong with these transformations, and they are available within R, but a simpler approach is often preferable and is likely to produce a model that is easier to interpret.

The major difficulty with modelling proportion data is that the responses are **strictly bounded**. There is no way that the percentage dying can be greater than 100% or less than 0%. However, if we use simple techniques like regression or analysis of covariance, then the fitted model could quite easily predict negative values or values greater than 100%, especially if the variance was high and many of the data were close to 0 or close to 100%.

The **logistic** curve is commonly used to describe data on proportions because, unlike the straight-line model, it asymptotes at 0 and 1 so that negative proportions – and responses of more than 100% cannot be predicted. Throughout this discussion we shall use p to describe the proportion of individuals observed to respond in a given way. Because much of their jargon was derived from the theory of gambling, statisticians call these **successes** although, to a demographer measuring death rates this may seem somewhat macabre. The individuals that respond in other ways (the statistician's **failures**) are therefore $(1 - p)$ and we shall call the proportion of failures q. The third variable is the size of the sample, n, from which p was estimated (it is the binomial denominator, and the statistician's **number of attempts**).

An important point about the binomial distribution is that the variance is not constant. In fact, the variance of a binomial distribution with mean $= np$ is:

$$s^2 = npq$$

so that the variance changes with the mean like this:

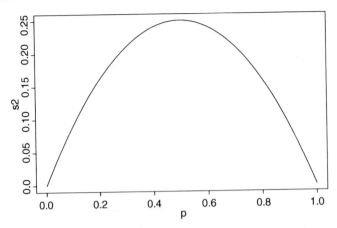

The variance is low when p is very high or very low, and the variance is greatest when $p = q = 0.5$. As p gets smaller, so the binomial distribution gets closer and closer to the Poisson distribution. You can see why this is so by considering the formula for the variance of the binomial (above). Remember that for the Poisson, the variance is equal to the mean: $s^2 = np$. Now, as p gets smaller, so q gets closer and closer to 1, so the variance of the binomial converges to the mean:

$$s^2 = npq \approx np \qquad (q \approx 1).$$

Odds

The logistic model for p as a function of x looks like this:

$$p = \frac{e^{(a+bx)}}{1 + e^{(a+bx)}},$$

and there are no prizes for realizing that the model is not linear; but if $x = -\infty$, then $p = 0$ and if $x = +\infty$ then $p = 1$ so the model is strictly bounded. When $x = 0$ then $p = \exp(a)/[(1 + \exp(a)]$. The trick of linearizing the logistic actually involves a very simple transformation. You may have come across the way in which bookmakers specify probabilities by quoting the **odds** against a particular horse winning a race (they might give odds of 2 to 1 on a reasonably good horse or 25 to 1 on an outsider). This is a rather different way of presenting information on probabilities than scientists are used to dealing with. Thus, where the scientist might state a proportion as 0.666 (2 out of 3), the bookmaker would give odds of 2 to 1 (2 successes to 1 failure). In symbols, this is the difference between the scientist stating the probability p, and the bookmaker stating the odds, p/q. Now if we take the **odds** p/q and substitute this into the formula for the logistic, we get:

$$\frac{p}{q} = \frac{e^{(a+bx)}}{1 + e^{(a+bx)}} \left[1 - \frac{e^{(a+bx)}}{1 + e^{(a+bx)}} \right]^{-1}$$

which looks awful. But a little algebra shows that:

$$\frac{p}{q} = \frac{e^{(a+bx)}}{1+e^{(a+bx)}} \left[\frac{1}{1+e^{(a+bx)}} \right]^{-1} = e^{(a+bx)}.$$

Now, taking natural logs, and recalling that $\ln(e^x) = x$ will simplify matters even further, so that

$$\ln\left(\frac{p}{q}\right) = a + bx.$$

This gives a **linear predictor**, $a + bx$, not for p but for the **logit** transformation of p, namely $\ln(p/q)$. In the jargon of R, the logit is the **link function** relating the linear predictor to the value of p.

Here is p as a function of x (left panel) and logit(p) as a function of x (right panel) for the logistic with $a = 0.2$ and $b = 0.1$:

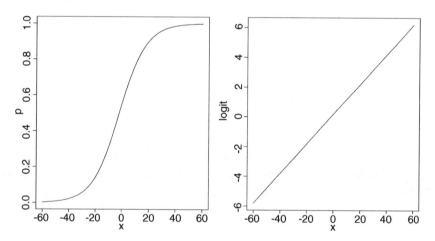

You might ask at this stage 'why not simply do a linear regression of $\ln(p/q)$ against the explanatory x-variable'? R has three great advantages here:

- it allows for the non-constant binomial variance;
- it deals with the fact that logits for p's near 0 or 1 are infinite;
- it allows for differences between the sample sizes by weighted regression.

Overdispersion and Hypothesis Testing

All the different statistical procedures that we have met in earlier chapters can also be used with data on proportions. Factorial analysis of variance, multiple regression, and a variety of models in which different regression lines are fit in each of several levels of one or more factors, can be carried out. The only difference is that we assess the significance of terms on the basis of chi-squared; the increase in scaled deviance that results from removal of the term from the current model.

The important point to bear in mind is that hypothesis testing with binomial errors is less clear-cut than with normal errors. While the chi-squared approximation for changes in scaled deviance is reasonable for large samples (i.e. bigger than about 30), it is poorer with small samples. Most worrisome is the fact that the degree to which the approximation is satisfactory is itself unknown. This means that considerable care must be exercised in the interpretation of tests of hypotheses on parameters, especially when the parameters are marginally significant or when they explain a very small fraction of the total deviance. With binomial or Poisson errors we cannot hope to provide exact p-values for our tests of hypotheses.

As with Poisson errors, we need to address the question of overdispersion (see Chapter 13). When we have obtained the minimal adequate model, **the residual scaled deviance should be roughly equal to the residual degrees of freedom**. When the residual deviance is larger than the residual degrees of freedom there are two possibilities: either the model is mis-specified, or the probability of success, p, is not constant within a given treatment level. The effect of randomly varying p is to increase the binomial variance from npq to

$$s^2 = npq + n(n-1)\sigma^2$$

leading to a large residual deviance. This occurs even for models that would fit well if the random variation were correctly specified.

One simple solution is to assume that the variance is not npq but $npqs$, where s is an unknown *scale parameter* $(s > 1)$. We obtain an estimate of the scale parameter by dividing the Pearson chi-square by the degrees of freedom, and use this estimate of s to compare the resulting scaled deviances. To accomplish this, we use family = quasibinomial rather than family = binomial when there is overdispersion.

The most important points to emphasize in modelling with binomial errors are as follows.

- Create a two-column object for the response, using cbind to join together the two vectors containing the counts of success and failure.

- Check for overdispersion (residual deviance > residual degrees of freedom), and correct for it by using family = quasibinomial rather than binomial if necessary.

- Remember that you do not obtain exact p-values with binomial errors; the chi-squared approximations are sound for large samples, but small samples may present a problem.

- The fitted values are counts, like the response variable.

- The linear predictor is in logits (the log of the odds $= \ln(p/q)$).

- You can back transform from logits (z) to proportions (p) by $p = 1/(1 + 1/\exp(z))$.

Applications

You can do as many kinds of modelling in a glm as in a linear model. Here we show examples of:

- regression with binomial errors (continuous explanatory variables),

- analysis of deviance with binomial errors (categorical explanatory variables),

- analysis of covariance with binomial errors (both kinds of explanatory variables).

Logistic Regression with Binomial Errors

This example concerns sex ratios in insects (the proportion of all individuals that are males). In the species in question, it has been observed that the sex ratio is highly variable, and an experiment was set up to see whether population density was involved in determining the fraction of males.

```
numbers < -read.table("c:\\temp\\sexratio.txt",header = T)
numbers
```

	density	females	males
1	1	1	0
2	4	3	1
3	10	7	3
4	22	18	4
5	55	22	33
6	121	41	80
7	210	52	158
8	444	79	365

It certainly looks as if there are proportionally more males at high density, but we should plot the data as proportions to see this more clearly:

```
attach(numbers)
par(mfrow = c(1,2))
p < -males/(males + females)
plot(density,p,ylab = "Proportion male")
plot(log(density),p,ylab = "Proportion male")
```

Evidently, a logarithmic transformation of the explanatory variable is likely to improve the model fit. We shall see in a moment.

The question is: 'does increasing population density lead to a significant increase in the proportion of males in the population'? or, more briefly, 'is the sex ratio density dependent'? – it certainly looks from the plot as if it is.

The response variable is a matched pair of counts that we wish to analyse as proportion data using a glm with binomial errors. First we bind together the vectors of male and female counts into a single object that will be the response in our analysis:

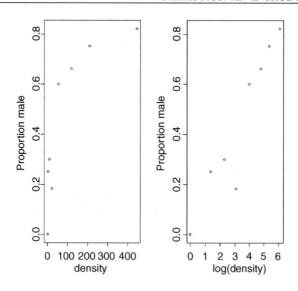

y < -cbind(males,females)

This means that y will be interpreted in the model as the proportion of all individuals that were male. The model is specified like this:

model < -glm(y ~ density,binomial)

This says that the object called 'model' gets a generalized linear model (glm) in which y (the sex ratio) is modelled as a function of a single continuous explanatory variable called density, using an error distribution from the family = binomial. The output looks like this:

summary(model)

```
Coefficients:
                Estimate    Std. Error   z value   Pr(>|z|)
(Intercept)     0.0807368   0.1550355    0.521     0.603
density         0.0035101   0.0005115    6.862     6.8e-12   ***

    Null deviance: 71.159 on 7 degrees of freedom
Residual deviance: 22.091 on 6 degrees of freedom
AIC: 54.618
```

The model table looks just as it would for a straightforward regression. The first parameter is the intercept and the second is the slope of the graph of sex ratio against population density. The slope is highly significantly steeper than zero (proportionately more males at higher population density: $p = 6.8 \ 10^{-12}$). We can see if log transformation of the explanatory variable reduces the residual deviance below 22.091

```
model < -glm(y ~ log(density),binomial)
summary(model)
```

```
Coefficients:
                 Estimate    Std. Error    z value    Pr(>|z|)
(Intercept)      -2.65927       0.48754     -5.454    4.91e-08    ***
log(density)      0.69410       0.09055      7.665    1.79e-14    ***
```

```
(Dispersion parameter for binomial family taken to be 1)
```

```
    Null deviance: 71.1593 on 7 degrees of freedom
Residual deviance:  5.6739 on 6 degrees of freedom
AIC: 38.201
```

This is a big improvement, so we shall adopt it. There is a technical point here, too. In a glm like this, it is assumed that the residual deviance is the same as the residual degrees of freedom. If the residual deviance is larger than the residual degrees of freedom, this is called overdispersion. It means that there is extra unexplained variation, over and above the binomial variance assumed by the model specification. In the model with log(density) there is no evidence of overdispersion (residual deviance = 5.67 on 6 d.f.), whereas the lack of fit introduced by the curvature in our first model caused substantial overdispersion (residual deviance = 22.09 on 6 d.f.).

Model checking involves the use of plot(model). As you will see, there is no pattern in the residuals against the fitted values, and the normal plot is reasonably linear. Point number 4 is highly influential (it has a big value of Cook's distance), but the model is still significant with this point omitted.

We conclude that the proportion of animals that are males increases significantly with increasing density, and that the logistic model is linearized by logarithmic transformation of the explanatory variable (population density). We finish by drawing the fitted line though the scatter plot:

```
xv<-seq(0,6,0.05)
plot(log(density),p,ylab = "Proportion male")
lines(xv,predict(model,list(density = exp(xv)),type = "response"))
```

Note the use of type = "response" to back-transform from the logit scale to the S-shaped proportion scale.

Proportion Data with Categorical Explanatory Variables

This example concerns the germination of seeds of two genotypes of the parasitic plant *Orobanche* and two extracts from host plants (bean and cucumber) that were used to stimulate germination. It is a two-way factorial analysis of deviance.

```
germination < -read.table("c:\\temp\\germination.txt",header = T)
attach(germination)
names(germination)
```

```
[1] "count"    "sample"    "Orobanche"    "extract"
```

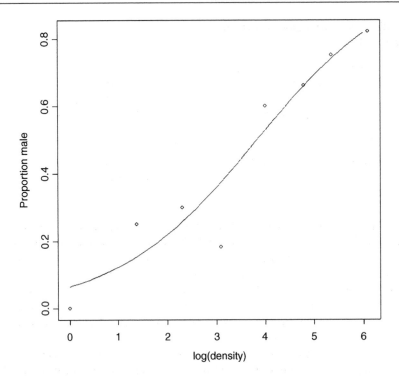

Count is the number of seeds that germinated out of a batch of size = sample. So the number that didn't germinate is sample − count, and we construct the response vector like this

y < -cbind(count, sample-count)

Each of the categorical explanatory variables has two levels

levels(Orobanche)

```
[1] "a73"  "a75"
```

levels(extract)

```
[1] "bean"  "cucumber"
```

We want to test the hypothesis that there is no interaction between *Orobanche* genotype ('a73' or 'a75') and plant extract ('bean' or 'cucumber') on the germination rate of the seeds. This requires a factorial analysis using the asterisk * operator like this

model < -glm(y ~ Orobanche * extract, binomial)

summary(model)

```
Coefficients:
                              Estimate   Std. Error   z value  Pr(>|z|)
(Intercept)                    -0.4122      0.1842     -2.238    0.0252  *
Orobanchea75                   -0.1459      0.2232     -0.654    0.5132
extractcucumber                 0.5401      0.2498      2.162    0.0306  *
Orobanchea75:extractcucumber    0.7781      0.3064      2.539    0.0111  *

(Dispersion parameter for binomial family taken to be 1)

    Null deviance: 98.719 on 20 degrees of freedom
Residual deviance: 33.278 on 17 degrees of freedom
AIC: 117.87
```

At first glance, it looks as if there is a highly significant interaction ($p = 0.0111$), but we need to check that the model is sound. The first thing to check for is overdispersion. The residual deviance is 33.278 on 17 d.f. so the model is quite badly overdispersed:

33.279/17

[1]   1.957588

The overdispersion factor is almost 2. The simplest way to take this into account is to use what is called an 'empirical scale parameter' to reflect the fact that the errors are not binomial as we assumed, but were larger than this (overdispersed) by a factor of 1.9576. We re-fit the model using quasibinomial to account for the overdispersion.

model < -glm(y ~ Orobanche * extract, quasibinomial)

Then we use update to remove the interaction term in the normal way.

model2 < -update(model, ~ . − Orobanche:extract)

The only difference is that we use an *F*-test instead of a Chi-square test to compare the original and simplified models:

anova(model,model2,test = "F")

```
Analysis of Deviance Table

Model 1: y~Orobanche * extract
Model 2: y~Orobanche + extract

  Resid. Df  Resid. Dev   Df  Deviance       F    Pr(>F)
1     17        33.278
2     18        39.686    -1    -6.408   3.4419   0.08099.
```

Now you see that the interaction is not significant ($p = 0.081$). There is no compelling evidence that different genotypes of *Orobanche* respond differently to the two plant extracts. The next step is to see if any further model simplification is possible.

anova(model2,test="F")

```
Analysis of Deviance Table

Model: quasibinomial, link: logit
Response: y

            Df   Deviance   Resid. Df   Resid. Dev        F      Pr(>F)
NULL                            20        98.719
Orobanche    1     2.544       19        96.175    1.1954     0.2887
extract      1    56.489       18        39.686   26.5412   6.692e-05  ***
```

There is a highly significant difference between the two plant extracts on germination rate, but it is not obvious that we need to keep *Orobanche* genotype in the model so we try removing it.

model3 <-update(model2, ~. –Orobanche)
anova(model2,model3,test="F")

```
Analysis of Deviance Table

Model 1: y~Orobanche + extract
Model 2: y~extract
     Resid. Df    Resid. Dev    Df   Deviance      F    Pr(>F)
1         18          39.686
2         19          42.751    -1    -3.065   1.4401   0.2457
```

There is no justification for retaining *Orobanche* in the model. So the minimal adequate model contains just two parameters:

coef(model3)

```
(Intercept)       extract
 -0.5121761      1.0574031
```

What, exactly, do these two numbers mean? Remember that the coefficients are from the linear predictor. They are on the transformed scale, so because we are using binomial errors, they are in logits $[\ln(p/(1-p))]$. To turn them into the germination rates for the two plant extracts requires a little calculation. To go from a logit x to a proportion p, you need to do the following sum

$$p = \frac{1}{1+\frac{1}{e^x}}.$$

So our first x value is -0.5122 and we calculate

1/(1 + 1/(exp(-0.5122)))

```
[1]  0.3746779
```

This says that the mean germination rate of the seeds with the first plant extract was 37%. What about the parameter for extract (1.057). Remember that with categorical explanatory variables **the parameter values are differences between means**. So to get the second germination rate we **add 1.057 to the intercept** before back-transforming:

1/(1 + 1/(exp(−0.5122 + 1.0574)))

```
[ 1]  0.6330212
```

This says that the germination rate was nearly twice as great (63%) with the second plant extract (cucumber). Obviously we want to generalize this process, and also to speed up the calculations of the estimated mean proportions. We can use predict to help here, because type = "response" makes predictions on the back-transformed scale automatically:

tapply(predict(model3,type = "response"),extract,mean)

```
      bean     cucumber
0.3746835   0.6330275
```

It is interesting to compare these figures with the averages of the raw proportions. First we need to calculate the proportion germinating, *p*, in each sample

p < -count/sample

then we can find the average the germination rates for each extract

tapply(p,extract,mean)

```
      bean     cucumber
0.3487189   0.6031824
```

You see that this gives different answers. Not too different in this case, it's true, but different none the less. The correct way to average proportion data is to add up the total counts for the different levels of abstract, and only then to turn them into proportions:

tapply(count,extract,sum)

```
  bean   cucumber
  148      276
```

This means that 148 seeds germinated with bean extract and 276 with cucumber, but how many seeds were involved in each case?

tapply(sample,extract,sum)

```
  bean   cucumber
  395      436
```

This means that 395 seeds were treated with bean extract and 436 seeds were treated with cucumber. So the answers we want are 148/395 and 276/436 (i.e. the correct mean proportions). We automate the calculation like this:

```
as.vector(tapply(count,extract,sum))/as.vector(tapply(sample,extract,sum))
```

```
[1] 0.3746835  0.6330275
```

These are the correct mean proportions that were produced by glm. The moral here is that **you calculate the average of proportions by using total counts and total samples and not by averaging the raw proportions**.

To summarize this analysis:

- make a two-column response vector containing the successes and failures,

- use glm with family = binomial (you don't need to include 'family='),

- fit the maximal model (in this case it had four parameters),

- test for overdispersion,

- if, as here, you find overdispersion then use quasibinomial rather than binomial errors,

- begin model simplification by removing the interaction term,

- this was non-significant once we had adjusted for overdispersion,

- try removing main effects (we didn't need *Orobanche* genotype in the model),

- use plot to obtain your model-checking diagnostics,

- back transform using predict with the option type = "response" to obtain means.

Analysis of Covariance with Binomial Data

This example concerns flowering in five varieties of perennial plants. Replicated individuals in a fully randomized design were sprayed with one of six doses of a controlled mixture of growth promoters. After 6 weeks, plants were scored as flowering or not flowering. The count of flowering individuals forms the response variable. This is an Ancova because we have both continuous (dose) and categorical (variety) explanatory variables. We use logistic regression because the response variable is a count (flowered) that can be expressed as a proportion (flowered/number).

```
props < -read.table("c:\\temp\\flowering.txt",header = T)
attach(props)
names(props)
```

```
[1] "flowered"    "number"    "dose"    "variety"
```

```
y < -cbind(flowered,number-flowered)
pf < -flowered/number
```

```
pfc <-split(pf,variety)
dc <-split(dose,variety)
```

```
plot(dose,pf,type = "n",ylab = "Proportion flowered")
points(dc[[1]],pfc[[1]],pch = 16)
points(dc[[2]],pfc[[2]],pch = 1)
points(dc[[3]],pfc[[3]],pch = 17)
points(dc[[4]],pfc[[4]],pch = 2)
points(dc[[5]],pfc[[5]],pch = 3)
```

There is clearly a substantial difference between the plant varieties in their response to the flowering stimulant if cut. The modelling proceeds in the normal way. We begin by fitting the maximal model with different slopes and intercepts for each variety (estimating ten parameters in all):

```
model1 <-glm(y ~ dose*variety,binomial)
summary(model1)
```

```
Coefficients:
```

	Estimate	Std. Error	z value	Pr(>|z|)	
(Intercept)	-4.591189	1.021236	-4.496	6.93e-06	***
dose	0.412564	0.099107	4.163	3.14e-05	***
varietyB	3.061504	1.082866	2.827	0.004695	**
varietyC	1.232022	1.178527	1.045	0.295842	
varietyD	3.174594	1.064689	2.982	0.002866	**
varietyE	-0.715041	1.537320	-0.465	0.641844	
dose:varietyB	-0.342767	0.101188	-3.387	0.000706	***
dose:varietyC	-0.230334	0.105826	-2.177	0.029515	*
dose:varietyD	-0.304762	0.101374	-3.006	0.002644	**
dose:varietyE	-0.006443	0.131786	-0.049	0.961006	

```
(Dispersion parameter for binomial family taken to be 1)

    Null deviance: 303.350 on 29 degrees of freedom
Residual deviance: 51.083 on 20 degrees of freedom
AIC: 123.55
```

The models exhibits substantial overdispersion, but this is probably due to poor model selection rather than extra, unmeasured variability. Here are the mean proportion flowered at each dose for each variety:

```
p <-flowered/number
tapply(p,list(dose,variety),mean)
```

	A	B	C	D	E
0	0.0000000	0.08333333	0.00000000	0.06666667	0.0000000
1	0.0000000	0.00000000	0.14285714	0.11111111	0.0000000

4	0.0000000	0.20000000	0.06666667	0.15789474	0.0000000
8	0.4000000	0.50000000	0.17647059	0.53571429	0.1578947
16	0.8181818	0.90000000	0.25000000	0.73076923	0.7500000
32	1.0000000	0.50000000	1.00000000	0.77777778	1.0000000

There are several ways to plot the five different curves on the scatterplot, but perhaps the simplest is to fit the regression model separately for each variety (see http://www.imperial.ac.uk/bio/research/crawley/statistics):

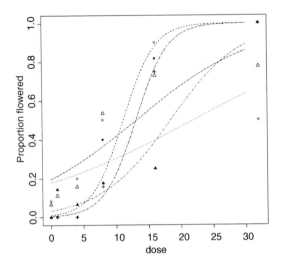

As you can see, the model is reasonable for two of the genotypes (A and E, represented by open and solid diamonds respectively), moderate for one genotype (C, solid triangles) but poor for two of them: B (open circles) and D (the open triangles). For both of the latter, the model overestimates the proportion flowering at zero dose, and for genotype B there seems to be some inhibition of flowering at the highest dose because the graph falls from 90% flowering at dose 16 to just 50% at dose 32. Variety D appears to be asymptoting at less than 100% flowering. These failures of the model focus attention for future work.

The moral is that just because we have proportion data, does not mean that the data will necessarily be well described by the logistic. For instance, in order to describe the response of genotype B, the model would need to have a hump, rather than to asymptote at $p = 1$ for large doses.

15

Death and Failure Data

Time-to-death data, and data on failure times, are often encountered in statistical model-ling. The main problem is that the variance in such data is almost always non-constant, and so standard methods are inappropriate. If the errors are gamma distributed, then the **variance is proportional to the square of the mean** (remember that with Poisson errors, the variance is equal to the mean). It is straightforward to deal with such data using a glm with Gamma errors.

This case study has 50 replicates in each of three treatments: an untreated control, low dosage and high dosage of a novel cancer treatment. The response is the age at death for the rats (expressed as an integer number of months):

```
mortality <-read.table("c:\\temp\\deaths.txt",header = T)
attach(mortality)
names(mortality)
```

```
[ 1]   "death"      "treatment"
```

```
tapply(death,treatment,mean)
```

```
control      high       low
  3.46      6.88      4.70
```

The animals receiving the high dose lived roughly twice as long as the untreated controls. The low dose increased life expectancy by more than 35%. The variance in age at death, however, is not constant

```
tapply(death,treatment,var)
```

```
  control           high              low
0.4167347      2.4751020      0.8265306
```

The variance is much greater for the longer-lived individuals, so we should not use standard statistical models which assume constant variance and Normal errors. However, we can use a generalized linear model with Gamma errors:

Statistics: An Introduction using R   M. J. Crawley
© 2005 John Wiley & Sons, Ltd   ISBNs: 0-470-02298-1 (PBK); 0-470-02297-3 (PPC)

```
model < -glm(death ~ treatment,Gamma)
summary(model)
```

```
Coefficients:
                  Estimate   Std. Error   t value   Pr(>|t|)
(Intercept)       0.289017   0.008304     34.804    < 2e-16  ***
treatmenthigh    -0.143669   0.009293    -15.461    < 2e-16  ***
treatmentlow     -0.076251   0.010311     -7.395    9.93e-12 ***
```

(Dispersion parameter for Gamma family taken to be 0.04136633)

```
    Null deviance: 17.7190 on 149 degrees of freedom
Residual deviance: 5.8337 on 147 degrees of freedom
AIC: 413.52
```

The link function with Gamma errors is the reciprocal so that is why the parameter for the high dose appears as a negative term in the summary table; the mean value for high dose is calculated as $0.289 - 0.1437 = 0.1453$, and $1/0.1453 = 6.882$. Checking the model using plot(model) shows that it is reasonably well-behaved (you might like to compare the behaviour of lm(death ~ treatment)). We conclude that all three treatment levels are significantly different from one another.

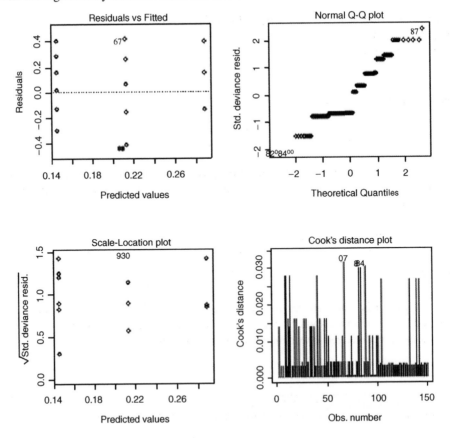

A common difficulty with data on time at death is that some (or even many) of the individuals do not die during the trial, so their age at death remains unknown (they might recover, they might leave the trial, or the experiment might end before they die). These individuals are said to be **censored**. Censoring makes the analysis much more complicated, because the censored individuals provide some information (we know the age at which they were last seen alive) but the data are of a different type from the information on age at death which is the response variable in the main analysis. There is a whole field of statistical modelling for such data and it is called **survival analysis**.

Survival Analysis with Censoring

The next example comes from a study of mortality in 150 wild male sheep. There were three experimental **groups**, and the animals were followed for 50 months. The groups were treated with three different medicines against their gut parasites: group A received a bolus with a high dose of worm-killer, group B received a low dose, and group C received the placebo (a bolus with no worm-killing contents). The initial body mass of each individual (**weight**) was recorded as a covariate. The month in which each animal died (**death**) was recorded, and animals which survived up to the 50th month (the end of the study) were recorded as being censored (for them, the censoring indicator **status** = 0, whereas the animals that died all have **status** = 1).

```
library(survival)
sheep <-read.table("c:\\temp\\sheep.txt",header = T)
attach(sheep)
names(sheep)
```

```
[ 1] "death"  "status"  "weight"  "group"
```

The overall survivorship curves for the three groups of animals are obtained like this:

```
plot(survfit(Surv(death,status) ~ group),lty = c(1,3,5),xlab = "Age at death (months)")
```

The crosses + at the end of the survivorship curves for groups A and B indicate that there was censoring in these groups (not all of the individuals were dead at the end of the experiment). Parametric regression in survival models uses the survreg function, for which you can specify a wide range of different error distributions. Here we use the exponential distribution for the purposes of demonstration (you can chose from dist = "extreme", "logistic", "gaussian" or "exponential" and from link = "log" or "identity"). We fit the full analysis of covariance model to begin with:

```
model <-survreg(Surv(death,status) ~ weight*group,dist = "exponential")
summary(model)
```

```
Call:
survreg(formula=Surv(death,status)~weight * group, dist =
"exponential")
```

	Value	Std. Error	z	p
(Intercept)	3.8702	0.3854	10.041	1.00e-23
weight	-0.0803	0.0659	-1.219	2.23e-01

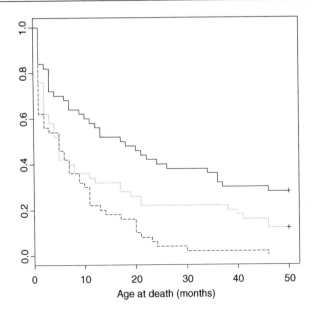

groupB	-0.8853	0.4508	-1.964	4.95e-02
groupC	-1.7804	0.4386	-4.059	4.92e-05
weight:groupB	0.0643	0.0674	0.954	3.40e-01
weight:groupC	0.0796	0.0674	1.180	2.38e-01

Scale fixed at 1

Exponential distribution
Loglik(model)= -480.6 Loglik(intercept only)= -502.1
          Chisq= 43.11 on 5 degrees of freedom,   p= 3.5e-08
Number of Newton-Raphson Iterations: 4

   Model simplification proceeds in the normal way. You could use **update**, but here (for variety only) we re-fit progressively simpler models and test them using **anova**. First we take out the different slopes for each group:

```
model2 < -survreg(Surv(death,status) ~ weight + group,dist = "exponential")
anova(model,model2,test = "Chi")
```

	Terms	Resid. Df	-2*LL	Test	Df	Deviance	P(>|Chi|)
1	weight * group	144	961.1800		NA	NA	NA
2	weight + group	146	962.9411	-weight:group	-2	-1.761142	0.4145462

   The interaction is not significant so we leave it out and try deleting weight:

```
model3 < -survreg(Surv(death,status) ~ group,dist = "exponential")
anova(model2,model3,test = "Chi")
```

	Terms	Resid. Df	-2*LL	Test	Df	Deviance	P(>|Chi|)
1	weight + group	146	962.9411		NA	NA	NA
2	group	147	963.9393	-weight	-1	-0.9981333	0.3177626

This is not significant, so we leave it out and try deleting group:

```
model4 < -survreg(Surv(death,status) ~ 1,dist = "exponential")
anova(model3,model4,test = "Chi")
```

	Terms	Resid. Df	-2*LL	Test Df	Deviance	P(>|Chi|)
1	group	147	963.9393	NA	NA	NA
2	1	149	1004.2865	-2	-40.34721	1.732661e-09

This is highly significant, so we add it back. The minimal adequate model is model 3 with the three-level factor **group**, but there is no evidence that initial body **weight** had any influence on survival.

```
summary(model3)
```

```
Call:
survreg(formula = Surv(death, status)~group, dist = "exponential")
```

	Value	Std. Error	z	p
(Intercept)	3.467	0.167	20.80	3.91e-96
groupB	-0.671	0.225	-2.99	2.83e-03
groupC	-1.386	0.219	-6.34	2.32e-10

```
Scale fixed at 1

Exponential distribution
Loglik(model) = -482    Loglik(intercept only) = -502.1
          Chisq= 40.35 on 2 degrees of freedom, p= 1.7e-09
Number of Newton-Raphson Iterations: 4
n= 150
```

We need to retain all three groups (group B is significantly different from both group A and group C).

It is straightforward to compare error distributions for the same model structure:

```
model3 < -survreg(Surv(death,status) ~ group,dist = "exponential")
model4 < -survreg(Surv(death,status) ~ group,dist = "extreme")
model5 < -survreg(Surv(death,status) ~ group,dist = "gaussian")
model6 < -survreg(Surv(death,status) ~ group,dist = "logistic")
anova(model3,model4,model5,model6)
```

	Terms	Resid. Df	-2*LL	Test Df	Deviance	P(>|Chi|)
1	group	147	963.9393	NA	NA	NA
2	group	146	1225.3512	=1	-261.411949	8.44789e-59
3	group	146	1178.6582	=0	46.692975	NaN
4	group	146	1173.9478	=0	4.710457	NaN

Our initial choice of exponential was clearly the best, giving much the lowest residual deviance (963.94).

You can immediately see the advantage of doing proper survival analysis when you compare the predicted mean ages at death from model 3 with the crude arithmetic averages of the raw data on age at death:

tapply(predict(model3,type = "response"),group,mean)

A	B	C
32.05555	16.38635	8.02

tapply(death,group,mean)

A	B	C
23.08	14.42	8.02

If there is no censoring (as in Group C, where all the individuals died) then the estimated mean ages at death are identical. However, when there is censoring, the arithmetic mean underestimates the age at death, and when the censoring is substantial (as in Group A) this underestimate is very large (23.08 *vs.* 32.06 months).

16

Binary Response Variable

Many statistical problems involve binary response variables. For example, we often classify individuals as

- dead or alive,

- occupied or empty,

- healthy or diseased,

- wilted or turgid,

- male or female,

- literate or illiterate,

- mature or immature,

- solvent or insolvent, or

- employed or unemployed.

It is interesting to understand the factors that are associated with an individual being in one class or the other. In a study of company insolvency, for instance, the data would consist of a list of measurements made on the insolvent companies (their age, size, turnover, location, management experience, workforce training and so on) and a similar list for the solvent companies. The question then becomes which, if any, of the explanatory variables increase the probability of an individual company being insolvent?

The response variable contains only 0s or 1s; for example, 0 to represent dead individuals and 1 to represent live ones. Thus, there is only a single column of numbers for the response, in contrast to proportion data where two vectors (successes and failures) were bound together to form the response (see Chapter 14). The way that R treats binary data is to assume that the 0s and 1s come from **a binomial trial with sample size 1**. If the probability that an individual is dead is p, then the probability of obtaining y (where y is

Statistics: An Introduction using R   M. J. Crawley
© 2005 John Wiley & Sons, Ltd   ISBNs: 0-470-02298-1 (PBK); 0-470-02297-3 (PPC)

either dead or alive, 0 or 1) is given by an abbreviated form of the binomial distribution with $n = 1$, known as the Bernoulli distribution:

$$P(y) = p^y(1 - p)^{(1-y)}.$$

The random variable y has a mean of p and a variance of $p(1 - p)$, and the objective is to determine how the explanatory variables influence the value of p. The trick for using binary response variables effectively is to know when it is worth using them, and when it is better to lump the successes and failures together and analyse the **total counts** of dead individuals, occupied patches, insolvent firms or whatever. The question you need to ask yourself is: **do I have unique values of one or more explanatory variables for each and every individual case?**

If the answer is 'yes', then analysis with a binary response variable is likely to be fruitful. If the answer is 'no', then there is nothing to be gained, and you should reduce your data by aggregating the counts to the resolution at which each count **does** have a unique set of explanatory variables. For example, suppose that all your explanatory variables were categorical such as gender (male or female), employment (employed or unemployed) and region (urban or rural). In this case there is nothing to be gained from analysis using a binary response variable because none of the individuals in the study have **unique** values of any of the explanatory variables. It might be worthwhile if you had each individual's body weight, for example, then you could ask the question 'when I control for gender and region, are heavy people more likely to be unemployed than light people?' In the absence of **unique** values for any explanatory variables, there are two useful options.

- Analyse the data as a contingency table using Poisson errors, with the count of the total number of individuals in each of the eight contingencies (2 x 2 x 2) as the response variable (see Chapter 13) in a dataframe with just eight rows.

- Decide which of your explanatory variables is the key (perhaps you are interested in gender differences), then express the data as proportions (the number of males and the number of females) and re-code the binary response as a count of a two-level factor. The analysis is now of proportion data (the proportion of all individuals that are female, for instance) using binomial errors (see Chapter 14).

If you **do** have unique measurements of one or more explanatory variables for each individual, these are likely to be continuous variables such as body weight, income, medical history, distance to the nuclear reprocessing plant, geographic isolation and so on. This being the case, successful analyses of binary response data tend to be multiple regression analyses or complex analyses of covariance, and you should consult Chapters 10 and 11 for details on model simplification and model criticism.

In order to carry out modelling on a binary response variable we take the following steps:

- create a single vector containing 0s and 1s as the response variable,

- use glm with family = binomial,

- you can change the link function from default logit to complementary log–log,

- fit the model in the usual way,

- test significance by deletion of terms from the maximal model, and compare the change in deviance with chi-square,

- note that there is no such thing as overdispersion with a binary response variable, and hence no need to change to using quasibinomial when the residual deviance is large.

Choice of link function is generally made by trying both links and selecting the link that gives the lowest deviance. The logit link that we used earlier is symmetric in p and q, but the complementary log–log link is asymmetric.

Incidence Functions

In this example, the response variable is called 'incidence'; a value of 1 means that an island was occupied by a particular species of bird, and 0 means that the bird did not breed there. The explanatory variables are the area of the island (km^2) and the isolation of the island (distance from the mainland, km).

```
island < -read.table("c:\\temp\\isolation.txt",header = T)
attach(island)
names(island)
```

```
[ 1]  "incidence"  "area"  "isolation"
```

There are two continuous explanatory variables, so the appropriate analysis is multiple regression. The response is binary, so we shall do logistic regression with binomial errors. We begin by fitting a complex model involving an interaction between isolation and area:

```
model1 < -glm(incidence ~ area*isolation,binomial)
```

then fit a simpler model with only main effects for isolation and area:

```
model2 < -glm(incidence ~ area + isolation,binomial)
```

then compare the two models using Anova:

```
anova(model1,model2,test = "Chi")
```

```
Analysis of Deviance Table
```

```
Model 1: incidence~area * isolation
Model 2: incidence~area + isolation
Resid.  Df  Resid. Dev  Df  Deviance  P(>|Chi|)
1        46    28.2517
2        47    28.4022  -1   -0.1504    0.6981
```

The simpler model is not significantly worse, so we accept this for the time being, and inspect the parameter estimates and standard errors:

summary(model2)

```
Call:
glm(formula = incidence~area + isolation, family = binomial)

Deviance Residuals:
     Min          1Q      Median          3Q         Max
 -1.8189     -0.3089      0.0490      0.3635      2.1192

Coefficients:
                 Estimate  Std. Error  z value  Pr(>|z|)
(Intercept)        6.6417      2.9218    2.273   0.02302   *
area               0.5807      0.2478    2.344   0.01909   *
isolation         -1.3719      0.4769   -2.877   0.00401   **

(Dispersion parameter for binomial family taken to be 1)

    Null deviance: 68.029 on 49 degrees of freedom
Residual deviance: 28.402 on 47 degrees of freedom
```

The estimates and their standard errors are in logits. Area has a significant positive effect (larger islands are more likely to be occupied), but isolation has a very strong negative effect (isolated islands are much less likely to be occupied). This is the minimal adequate model. We should plot the fitted model through the scatterplot of the data. It is much easier to do this for each variable separately, like this:

```
modela < -glm(incidence~area,binomial)
modeli < -glm(incidence~isolation,binomial)
par(mfrow = c(1,2))
xv < -seq(0,9,0.01)
yv < -predict(modela,list(area = xv),type = "response")
plot(area,incidence)
lines(xv,yv)
xv2 < -seq(0,10,0.1)
yv2 < -predict(modeli,list(isolation = xv2),type = "response")
plot(isolation,incidence)
lines(xv2,yv2)
```

This is all well and good, but it is very difficult to know how good the fit of the model is when the data are shown only as zeros or ones. It is sensible to compute one or more intermediate probabilities from the data, and to show these empirical estimates (ideally with their standard errors) on the plot in order to judge whether the fitted line is a reasonable description of the data.

For the purposes of demonstration, we take the central third of the data ranked by area and by isolation, calculate the mean proportion incidence, p, and add this to the plot

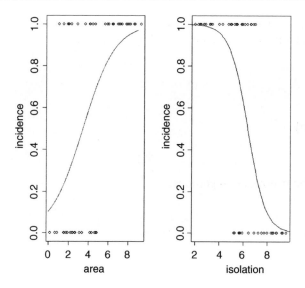

along with its standard error $\sqrt{p(1-p)/n}$. We use cut to obtain the central third of the data:

```
ac <-cut(area,3)
ic <-cut(isolation,3)
tapply(incidence,ac,sum)
```

(0.144,3.19]	(3.19,6.23]	(6.23,9.28]
7	8	14

```
tapply(incidence,ic,sum)
```

(2.02,4.54]	(4.54,7.06]	(7.06,9.58]
12	17	0

Note the convention for labelling intervals: (a, b] means include their left-hand endpoint, a, but not their right-hand one, b. Now count the number of cases in each interval using table

```
table(ac)
ac
```

(0.144,3.19]	(3.19,6.23]	(6.23,9.28]
21	15	14

```
table(ic)
ic
```

(2.02,4.54]	(4.54,7.06]	(7.06,9.58]
12	25	13

A sensible place to plot the mean probability associated with the central third of the explanatory variable is in the position defined by the median:

median(area)

```
[ 1]  4.1705
```

median(isolation)

```
[ 1]  5.8015
```

Next, calculate the two mean proportions:

8/15

```
[ 1]  0.5333333
```

17/25

```
[ 1]  0.68
```

and their two standard errors:

sqrt((8/15*7/15)/15)

```
[ 1]  0.1288122
```

sqrt((17/25*8/25)/25)

```
[ 1]  0.09329523
```

Finally, re-plot the two graphs, adding the empirical estimates and their error bars:

```
plot(area,incidence)
lines(xv,yv)
points(4.1705,0.5333,pch = 16)
lines(c(4.1705,4.1705),c(0.533333-0.1288,0.533333 + 0.1288))
plot(isolation,incidence)
lines(xv2,yv2)
points(5.8015,0.68,pch = 16)
lines(c(5.8015,5.8015),c(0.68-0.093,0.68 + 0.093))
```

This shows that the fit to the central third of the data is excellent for the relationship between incidence and isolation, but less good (although not significantly far out) for the relationship with area. With a large data set, you can compute more of these empirical estimates (say three, five, or seven of them) and this would enable you to test quite sensitively for model failure. This approach would not work, of course, if there was an interaction between area and isolation; then you would need to produce conditioning plots of incidence against area for different degrees of isolation.

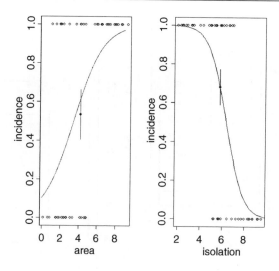

Ancova with a Binary Response Variable

In this example the binary response variable is parasite infection (infected or not) and the explanatory variables are weight and age (continuous) and gender (categorical). We begin with data inspection:

```
infection < -read.table("c:\\temp\\infection.txt",header = T)
attach(infection)
names(infection)
```

```
[ 1] "infected"  "age"  "weight"  "gender"
```

```
par(mfrow = c(1,2))
plot(infected,weight,xlab = "Infection",ylab = "Weight")
plot(infected,age,xlab = "Infection",ylab = "Age")
```

Infected individuals are substantially lighter than uninfected individuals and occur in a much narrower range of ages. To see the relationship between infection and gender (both categorical variables) we can use table:

```
table(infected,gender)
```

```
table(infected,gender)
                gender
infected        female       male
   absent          17          47
  present          11           6
```

which indicates that the infection is much more prevalent in females (11/28) than in males (6/53).

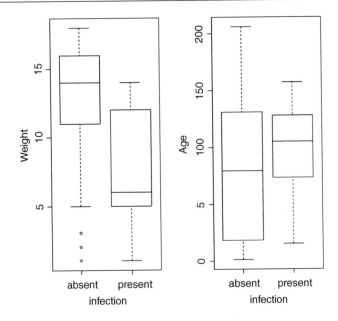

We begin, as usual, by fitting a maximal model with different slopes for each level of the categorical variable:

```
model < -glm(infected ~ age*weight*gender,family = binomial)
summary(model)
```

```
Coefficients:
                        Estimate   Std. Error   z value   Pr(>|z|)
(Intercept)            -0.109124     1.375388    -0.079      0.937
age                     0.024128     0.020874     1.156      0.248
weight                 -0.074156     0.147678    -0.502      0.616
gendermale             -5.969133     4.275952    -1.396      0.163
age:weight             -0.001977     0.002006    -0.985      0.325
age:gendermale          0.038086     0.041310     0.922      0.357
weight:gendermale       0.213835     0.342825     0.624      0.533
age:weight:gendermale  -0.001651     0.003417    -0.483      0.629
```

```
(Dispersion parameter for binomial family taken to be 1)

    Null deviance: 83.234 on 80 degrees of freedom
Residual deviance: 55.706 on 73 degrees of freedom
AIC: 71.706
```

It certainly does not look as if any of the high-order interactions are significant. Instead of using update and anova for model simplification, we can use step to compute the AIC for each term in turn (see p. 208).

```
model2 < -step(model)
```

```
Start: AIC= 71.71
```

First, it tests whether the three-way interaction is required

```
                      Df    Deviance        AIC
- age:weight:gender    1      55.943     69.943
⟨none⟩                        55.706     71.706

Step: AIC= 69.94
```

This causes a reduction in AIC of just $71.7 - 69.9 = 1.8$ and hence is not significant. Next, it looks at the three two-way interactions and decides which to delete first:

```
                    Df  Deviance      AIC
- weight:gender      1    56.122   68.122
- age:gender         1    57.828   69.828
⟨none⟩                    55.943   69.943
- age:weight         1    58.674   70.674

Step: AIC= 68.12
```

Only the removal of weight: gender causes a substantial reduction in AIC, so this interaction in deleted and the other two interactions are retained. Let's see if we would have been this lenient:

summary(model2)

```
Call:
glm(formula = infected~age + weight + gender + age:weight + age:gender,
    family = binomial)

Coefficients:
                 Estimate  Std. Error   z value    Pr(>|z|)
(Intercept)     -0.391572    1.264850    -0.310      0.7569
age              0.025764    0.014918     1.727      0.0842    .
weight          -0.036493    0.128907    -0.283      0.7771
gendermale      -3.743698    1.786011    -2.096      0.0361    *
age:weight      -0.002221    0.001365    -1.627      0.1037
age:gendermale   0.020464    0.015199     1.346      0.1782

(Dispersion parameter for binomial family taken to be 1)

    Null deviance: 83.234 on 80 degrees of freedom
Residual deviance: 56.122 on 75 degrees of freedom
AIC: 68.122
```

Neither of the two interactions retained by step would figure in our model ($p > 0.10$). We shall use update to simplify model2:

```
model3 < -update(model2, ~ .-age:weight)
anova(model2,model3,test = "Chi")
```

```
Analysis of Deviance Table
```

```
Model 1: infected~age + weight + gender + age:weight +age: gender
Model 2: infected~age + weight + gender + age:gender
Resid.  Df        Resid. Dev        Df      Deviance     P(>|Chi|)
1       75            56.122
2       76            58.899         -1       -2.777        0.096
```

so there is no really persuasive evidence of an age:weight term ($p = 0.096$)

```
model4 < -update(model2, ~ .-age:gender)
anova(model2,model4,test = "Chi")
```

Note that we are testing all the two-way interactions by deletion from the model that contains all two-way interactions (model 2): $p = 0.155$, so nothing there, then. What about the three main effects?

```
model5 < -glm(infected ~ age + weight + gender,family = binomial)
summary(model5)
```

```
Coefficients:
                 Estimate    Std. Error    z value    Pr(>|z|)
(Intercept)      0.609392     0.801303      0.761     0.446955
age              0.012649     0.006717      1.883     0.059654    .
weight          -0.227880     0.068138     -3.344     0.000825   ***
gendermale      -1.543151     0.681434     -2.265     0.023539    *
```

```
(Dispersion parameter for binomial family taken to be 1)

    Null deviance: 83.234 on 80 degrees of freedom
Residual deviance: 59.859 on 77 degrees of freedom
AIC: 67.859
```

Weight is highly significant, as we expected from the initial boxplot, gender is quite significant, and age is marginally significant. It is worth establishing whether there is any evidence of non-linearity in the response of infection to weight or age. We might begin by fitting quadratic terms for the two continuous explanatory variables:

```
model6 < -
glm(infected ~ age + weight + gender + I(weight^2) + I(age^2), family = binomial)
summary(model6)
```

```
Coefficients:
                 Estimate     Std. Error    z value    Pr(>|z|)
(Intercept)     -3.4469474     1.7825435    -1.934      0.0531    .
age              0.0829206     0.0355997     2.329      0.0198    *
weight           0.4465758     0.3355612     1.331      0.1832
```

```
gender              -1.2202485    0.7646071    -1.596       0.1105
I(weight^2)         -0.0415082    0.0208383    -1.992       0.0464    *
I(age^2)            -0.0004008    0.0001981    -2.023       0.0431    *
```

(Dispersion parameter for binomial family taken to be 1)

```
    Null deviance: 83.234 on 80 degrees of freedom
Residual deviance: 48.620 on 75 degrees of freedom
AIC: 60.62
```

Evidently, both relationships are significantly non-linear. It is worth looking at these non-linearities in more detail, to see if we can do better with other kinds of models (e.g. non-parametric smoothers, piece-wise linear models or step functions). A good start is often a gam (a generalized additive model) when we have continuous covariates:

```
library(mgcv)
model7 < -gam(infected ~ gender + s(age) + s(weight),family = binomial)
plot.gam(model7)
```

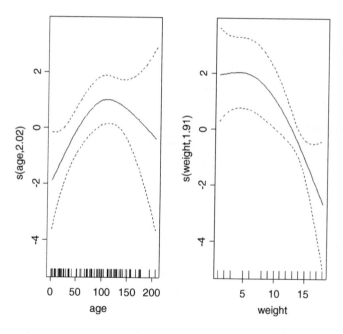

These non-parametric smoothers are excellent at showing the humped relationship between infection and age, and at highlighting the possibility of a threshold at weight ≈8 in the relationship between weight and infection. We can now return to a glm to incorporate these ideas. We shall fit age and age^2 as before, but try a piecewise linear fit for **weight**, estimating the threshold weight at a range of values (say 8–14) and selecting the threshold that gives the lowest residual deviance; this turns out to be a threshold = 12. The piecewise regression is specified by the term:

$$I((weight - 12) * (weight > 12))$$

The I ('as is') is necessary to stop the * as being evaluated as an interaction term in the model formula. What this expression says is 'regress infection on the value of weight − 12, but only do this when (weight > 12) is true'. Otherwise, assume that infection is independent of weight.

model8 < -glm(infected ∼ sex + age + I(age^2) + I((weight−12)* (weight > 12)),
  family = binomial)
summary(model8)

```
Coefficients:
                                         Estimate   Std. Error  z value   Pr(>|z|)
(Intercept)                            -2.7511452    1.3672006   -2.012     0.0442   *
gender male                            -1.2864620    0.7343309   -1.752     0.0798   .
age                                     0.0798630    0.0347926    2.295     0.0217   *
I(age^2)                               -0.0003892    0.0001953   -1.993     0.0463   *
I((weight - 12) * (weight > 12))       -1.3547080    0.5318043   -2.547     0.0109   *

(Dispersion parameter for binomial family taken to be 1)

     Null deviance: 83.234 on 80 degrees of freedom
Residual deviance: 48.687 on 76 degrees of freedom
AIC: 58.687
```

model9 < -update(model8, ∼.-gender)
anova(model8,model9,test = "Chi")
model10 < -update(model8, ∼.-I(age^2))
anova(model8,model10,test = "Chi")

The effect of gender on infection is not quite significant ($p = 0.071$ for a Chi-square test on deletion), so we leave it out. The quadratic term for age does not look highly significant here, but a deletion test gives $p = 0.011$, so we retain it. The minimal adequate model is therefore model9:

summary(model9)

```
Coefficients:
                                         Estimate   Std. Error  z value   Pr(>|z|)
(Intercept)                            -3.1207575    1.2663090   -2.464    0.01372   *
age                                     0.0765785    0.0323274    2.369    0.01784   *
I(age^2)                               -0.0003843    0.0001845   -2.082    0.03732   *
I((weight - 12) * (weight > 12))       -1.3511514    0.5112930   -2.643    0.00823   **

(Dispersion parameter for binomial family taken to be 1)

     Null deviance: 83.234 on 80 degrees of freedom
Residual deviance: 51.953 on 77 degrees of freedom
AIC: 59.953
```

We conclude there is a humped relationship between infection and age, and a threshold effect of weight on infection. The effect of gender is marginal, but might repay further investigation ($p = 0.071$).

Appendix 1: Fundamentals of the R language

R as a Calculator

Immediately to the right of the prompt symbol ' > ' on the command line is space in which you can perform a wide range of calculations. The arithmetic operators for addition, subtraction and division are $+$, $-$ and $/$ respectively, while $*$ means multiply ($\times$) and $\wedge$ means 'to the power'. The first operations to be carried out are powers, then multiplication and division, and finally addition and subtraction. You can overide this hierarchy of calculation by the use of brackets, so to calculate the cube root of 17×0.35 you would type

```
(17 * 0.35) ^ (1/3)
```
```
[ 1] 1.812059
```

There is a huge range of mathematical functions; the ones you will use most often are log (logarithm to base e), exp (natural antilog) and sqrt (square root).

```
log(10)
```
```
[ 1] 2.302585
```

```
exp(1)
```
```
[ 1] 2.718282
```

Negative powers are reciprocals, so x^{-1} is the same as $1/x$:

```
3^-1
```
```
[ 1] 0.3333333
```

```
1/3
```
```
[ 1] 0.3333333
```

Statistics: An Introduction using R   M. J. Crawley
© 2005 John Wiley & Sons, Ltd   ISBNs: 0-470-02298-1 (PBK); 0-470-02297-3 (PPC)

Assigning Values to Variables

Variables are assigned values in R using 'gets' $<-$ rather than the more familiar 'equals' $=$ sign. Gets is a composite operator made up of a 'less than' symbol $<$ and a minus sign $-$. To assign one value to a variable (creating a scalar) called x just write

x <-12.6

Now, whenever we use the variable x the value 12.6 is used in its place (until we change the value of x with another assignment). More usually in R we work with variables that contain many values (vectors). These can be assigned values in several different of ways. The simplest is to write down all the values separated by commas, and turn this into a vector using the **concatenate** function, c, like this:

y <-c(3,7,9,11)

If the vector of numbers was long, this would be tedious to type and difficult to proof read. Alternatively, you can type the numbers in at the keyboard during an R session, using the scan function like this:

z <-scan()

```
1: 8
2: 4
3: 7
4: 5
5:
Read 4 items
```

After typing each number (starting with 8 in this case), press the Return key, then number 4, Return key, and so on. To finish, type two successive Return keys. R responds by telling you how many values you have entered.

If the numbers you want to put into a vector form a regular sequence of some sort, then you can automate the procedure. Suppose you want the integer (whole) numbers 1 to 6 in a vector called a. You just type

a <-1:6

and R understands the colon operator : to mean 'a series of integers between'. Alternatively, your series might be in non-integer steps (say, in steps of 0.1) in which case you use the seq function. For decreasing series you specify negative values of the step size. The vector b contains six numbers stepped down from one half to zero:

b <-seq(0.5,0,-0.1)

Generating Repeats

The rep function replicates the object which is its first argument by the number of times specified in the second argument. Thus

```
rep("A",10)
```

```
[ 1]  "A"  "A"  "A"  "A"  "A"  "A"  "A"  "A"  "A"  "A"
```

produces ten copies of the character 'A'. The object to be repeated might be a series:

```
rep(1:6,2)
```

```
[ 1]  1  2  3  4  5  6  1  2  3  4  5  6
```

This says repeat the whole series, 1 to 6, twice. If we want the **elements** of a series to be repeated (rather than the series as a whole), then the second argument of the rep function needs to be a vector of the same length as the first argument. That sounds complicated, but suppose we wanted three 1's, then three 2's and so on up to three 6's, we would put

```
rep(1:6,rep(3,6))
```

```
[ 1]  1  1  1  2  2  2  3  3  3  4  4  4  5  5  5  6  6  6
```

The most complex case arises when we want to repeat each element of the first vector a different number of times. One symmetric case might be if we wanted one 1, two 2's, three 3's and so on. This is

```
rep(1:6,1:6)
```

```
[ 1]  1  2  2  3  3  3  4  4  4  4  5  5  5  5  5  6  6  6  6  6  6
```

but more generally, we would like to be able to specify each repeat separately. Here, the elements of the first vector c(4,7,1,5) are repeated by a number of times contained in the second vector c(3,2,5,2) (the two vectors must be the same length)

```
rep(c(4,7,1,5),c(3,2,5,2))
```

```
[ 1]  4  4  4  7  7  1  1  1  1  1  5  5
```

Generating Factor Levels

The function gl is very useful for generating levels of factors automatically. The arguments of the function are:

- 'up to', and
- 'with repeats of'.

Suppose you want to generate factor levels up to 5 with repeats of 3 you write

gl(5,3)

```
[1]        1  1   1   2   2   2   3   3   3   4   4   4   5   5   5
Levels:  1   2   3   4   5
```

By default this pattern is executed just once. If you want repeats of the whole pattern, you specify the total length of the object as the optional third argument. To get two repeats (i.e. total length = 30) you write

gl(5,3,30)

```
[1]        1  1  1 2  2  2 3 3 3 4 4  4 5 5 5 1 1  1 2 2  2 3 3  3 4 4 4  5 5 5
Levels:      1   2   3   4   5
```

Usefully, the function gl automatically declares the vector to be a factor. You can see this from:

is.factor(gl(5,3,30))

```
[1]  TRUE
```

Changing the Look of Graphics

The most likely changes required of R graphics are in the orientation and size of the labels for the x and y axes. Here are the defaults:

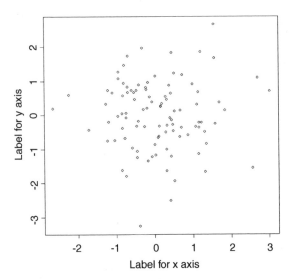

Many journals require that the y axis figures are vertically aligned like those on the x axis, rather than at right angles to the y axis. This is achieved by las = 1. If you want to make the text of the axis labels bigger (increase the font size), specify cex ('character expansion') for the labels, using cex.lab = 1.5

```
plot(rnorm(100),rnorm(100),ylab = "Label for y axis",
                      xlab = "Label for x axis",las = 1,cex.lab = 1.5)
```

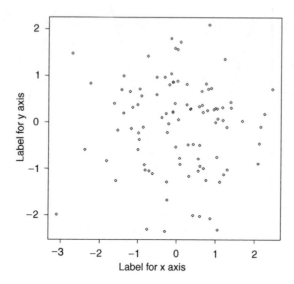

To see the full range of changes that you could make to graphics if you wanted to, it is worth spending a while browsing the help window on graphics parameters by typing

```
?par
```

You will be impressed by how much is being done for you, automatically, behind the scenes when you say plot(x,y).

To write on plots using more intricate mathematical symbols or Greek letters we use expression or substitute. Here are some examples of their use. First, we produce a plot of $\sin \phi$ against the phase angle ϕ on the range $-\pi$ to $+\pi$ radians:

```
x <- seq(-4, 4, len = 101)
plot(x,sin(x),type = "l",xaxt = "n",
   xlab = expression(paste("Phase Angle ",phi)),
   ylab = expression("sin "*phi))
axis(1, at = c(-pi, -pi/2, 0, pi/2, pi),
lab = expression(-pi, -pi/2, 0, pi/2, pi))
```

Note the use of xaxt = "n" to suppress the default labelling of the x axis, and the use of expression in the labels for the x and y axes to obtain mathematical symbols like phi or pi. The more intricate labels on the x axis are obtained by the axis function, specifying 1 (the x axis is the first axis), then using the at function to say where the labels and tick marks are to appear, and lab with expression to say what the labels are to be.

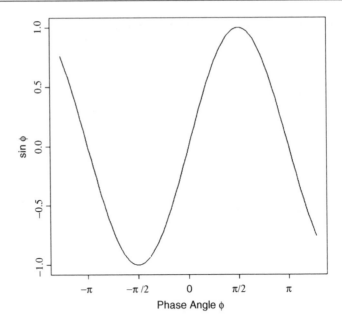

Suppose you wanted to add '$\chi^2 = 24.5$' to this graph at location $(-\pi/2, 0.5)$. You use text with substitute, like this:

text(-pi/2,0.5,substitute(chi^2 = = "24.5"))

Note the use of 'double equals' to print a single equals sign.

You can write quite complicated formulae on plots using paste to join together the elements of an equation: here is the density function of the Normal written on the plot at location $(\pi/2, -0.5)$:

text(pi/2, -0.5, expression(paste(frac(1, sigma*sqrt(2*pi)), " ",
        e^{frac(-(x-mu)^2, 2*sigma^2)})))

Note the use of frac to obtain individual fractions; the first argument is the text for the numerator, the second the text for the denominator. Most of the arithmetic operators have obvious formats $(+, -, /, *, \wedge$, etc.); the only non-intuitive symbol that is commonly used is 'plus or minus', $\pm$; this is written as % + −% like this:

text(pi/2,0,expression(hat(y) % + −% se))

Reading Data From a File

For real applications, the typical way to get numbers into a variable in R is to read them from a file that you created and error-checked earlier, perhaps in Excel. The vector may contain many thousands of numbers, and it would make no sense at all to type the numbers directly into R via the keyboard using scan. The R function that reads data from a file is called read.table and you use it like this:

```
dataframe < -read.table("c:\\directory\\filename.txt",header = T)
```

Note that the drive, the directory and the file name are enclosed in double quotes. The phrase header = T says that row 1 of your file contains the variable name(s). You must use 'double backslash' \\ to separate the drive, directory and file names.

Common things that can go wrong at this stage include:

- the file does not exist (e.g. you have mis-spelled the file name, or given it the wrong extension, .prn instead of .txt for instance);

- the file is not in the directory you specified, but somewhere else;

- the variable names in row 1 of your file have blank spaces between words in some of them (e.g. 'dry weight' or 'body mass') – in such a case, R would look for four columns of numbers ('dry', 'weight', 'body' and 'mass'), not two as intended and the solution is to use a dot instead of a blank in variable names (e.g. 'dry.weight' or 'body.mass');

- you have used blanks in the file to represent missing values (you should use NA).

To use variables contained within the dataframe, you need to use attach like this:

```
attach(dataframe)
```

and to see the names of the variables, and the order in which they appear use

```
names(dataframe)
```

When you have finished using a set of variables, tidy up by detaching the dataframe and removing (rm) any variables names that you have assigned using 'gets':

```
detach(dataframe)
rm(a,b,x,y,z)
```

Vector Functions in R

```
y < -c(5,7,7,8,2,5,6,6,7,5,8,3,4)
z < -13:1
```

Typical operations on vectors include summary statistics, using functions like mean, var, range, max, min, summary, IQR and fivenum.

```
mean(y)
```

```
[ 1] 5.615385
```

```
var(y)
```

```
[ 1] 3.423077
```

range(y)

```
[ 1]  2    8
```

Some functions (like range) produce more than one number as output (min and max in this case). A very powerful feature is that you can do arithmetic with entire vectors. The * operator performs vector multiplication. Since y and z are the same length, y*z gives a vector of the same length as y containing the point-wise products ($5 \times 13, 7 \times 12$, 7×11, etc.):

y*z

```
[ 1] 65  84   77   80   18   40   42   36   35   20   24   6   4
```

In R, if two vectors are not the same length, then the shorter vector is repeated as necessary, up to the length of the longer vector. You can see this with the expression

y*6

```
[ 1]  30   42   42   48   12   30   36   36   42   30   48   18   24
```

in which the scalar, 6, is repeated 13 times to match the length of y. Two vectors are joined together, the top of the second vector to the bottom of the first, using the concatenate function, c:

c(y,z)

```
[ 1]    5  7   7  8   2  5   6  6  7  5  8  3   4 13 12  11 10  9   8  7  6 5 4 3 2
[ 26]   1
```

Subscripts: Obtaining Parts of Vectors

Elements of vectors are addressed by subscripts which appear in square brackets []. The third element of y is extracted like this:

y[3]

```
[ 1]  7
```

the third to the seventh elements of y in sequence like this

y[3:7]

```
[ 1]  7   8    2    5    6
```

and the third, fifth, sixth and ninth elements of y like this

y[c(3,5,6,9)]

```
[ 1]  7  2  5  7
```

To drop an element from an array, you use negative subscripts. Here is the vector *y* without its first element

```
y[-1]
```

```
[ 1]  7   7   8   2   5   6   6   7   5   8   3   4
```

and here is a general way of dropping the last element of the array, without knowing in advance how long the array might be

```
y[-length(y)]
```

```
[ 1]  5   7   7   8   2   5   6   6   7   5   8   3
```

Subscripts as Logical Variables

Often you want to use some kind of logical condition to find a subset of the values in a vector. Suppose, for instance, that we wanted to know all the values of *y* that were bigger than 6? This could not be simpler in R: just state the logical condition as the subscript:

```
y[y > 6]
```

```
[ 1]  7   7   8   7   8
```

We might want to know the values of *z* for which $y > 6$. Again, this could not be simpler:

```
z[y > 6]
```

```
[ 1]  12   11   10   5   3
```

Suppose we wanted to extract all of the elements of *y* that were **not multiples of three**. If a number is a multiple of three then y%%3 ('y modulo 3') will be zero. The symbol for 'not equal' in R is ! = (exclamation, equals), so the way to extract the non-multiples of three is

```
y[y%%3! = 0]
```

```
[ 1]  5   7   7   8   2   5   7   5   8   4
```

and you see that all the threes and sixes have been removed from *y*.

Subscripts with Arrays

Begin by making a three-dimensional array containing the numbers 1 to 30, structured so that there are five rows and three columns in each of two tables. The first dimension refers to the number of rows, the second to the number of columns, and the third to the number of two-dimensional tables. Note that the numbers enter each table column-wise (rather than row-wise), and that the elements of the array are filled up through the dimensions of the array from left to right (rows then columns then tables):

```
A <- array(1:30, c(5,3,2) )
A
```

```
, ,  1

      [ ,1]      [ ,2]   [ ,3]
[1,]    1         6       11
[2,]    2         7       12
[3,]    3         8       13
[4,]    4         9       14
[5,]    5        10       15

, ,  2

      [ ,1]      [ ,2]   [ ,3]
[1,]    16        21      26
[2,]    17        22      27
[3,]    18        23      28
[4,]    19        24      29
[5,]    20        25      30
```

You might want to select only the second and third columns of A. Columns are the second (middle) subscript, so the first and third subscripts are left blank

```
A[,2:3,]
```

```
, ,  1

      [ ,1]   [ ,2]
[1,]     6      11
[2,]     7      12
[3,]     8      13
[4,]     9      14
[5,]    10      15

, ,  2

      [ ,1]   [ ,2]
[1,]    21      26
[2,]    22      27
[3,]    23      28
[4,]    24      29
[5,]    25      30
```

In another application you might want to take rows two to four of this reduced array (but keep both tables, so the last subscript is blank):

```
A[2:4,2:3,]
```

```
,  ,  1

          [ ,1]      [ ,2]
[ 1,]       7        12
[ 2,]       8        13
[ 3,]       9        14

,  ,  2

          [ ,1]      [ ,2]
[ 1,]      22         27
[ 2,]      23         28
[ 3,]      24         29
```

Finally, you might want only the second of the two tables of this reduced array, so all three subscripts are specified

A[2:4,2:3,2]

```
          [ ,1]      [ ,2]
[ 1,]      22         27
[ 2,]      23         28
[ 3,]      24         29
```

Subscripts with Lists

Vectors are subscripted like this [3], but lists are subscripted like this [[3]]. Understanding the distinction takes a good deal of practice. Here is a list called cars, with three elements to the list: make, capacity and colour

```
cars < -list(c("Toyota","Nissan","Honda"),
   c(1500,1800,1750),c("blue","red","black","silver"))
cars
```

```
[[ 1]]
[ 1] "Toyota"  "Nissan"  "Honda"

[[ 2]]
[ 1] 1500  1800  1750

[[ 3]]
[ 1] "blue"  "red"  "black"  "silver"
```

You need to understand the difference between cars[[3]] and cars[3]

cars[[3]]

```
[ 1] "blue"  "red"     "black"  "silver"
```

cars[3]

```
[[ 1]]
[ 1] "blue"  "red"   "black"   "silver"
```

The distinction is apparently rather subtle, but it is very important when you try to extract one element of the sub-list using subscripts (e.g. suppose we want to extract the second colour, red): double brackets works

```
cars[[3]][2]
```

```
[ 1]  "red"
```

but single brackets does not

```
cars[3][2]
```

```
[[ 1]]
NULL
```

Writing Functions in R

One of the outstanding features of R is the ease with which you can write your own functions. Here is a function to produce a summary of various measures of central tendency: we want to print the median, the arithmetic mean, the geometric mean and the harmonic mean of the numbers in a vector called x. Let's call the function central and define it like this:

```
central  <- function (x) {
gm < -exp(mean(log(x)))
hm < -1/mean(1/x)
cat("Median", median(x),"\n")
cat("Arithmetic mean",mean(x),"\n")
cat("Geometric mean",gm,"\n")
cat("Harmonic mean",hm,"\n")   }
```

Functions are created using function and the code is contained within 'curly brackets', {}. Lines of code are separated with a 'hard return'. There are no built-in functions for geometric mean or harmonic mean, so we have to write our own code on lines two and three. The rest of the code produces nicely formatted output. The function to do this is cat (it is like print but with control over format). The key point here is that to get a new line for the next bit of output, you must include "\n" at the end of your cat function. Now we can run the function to compare the different measures of central tendency for the data in y:

```
central(y)
```

```
Median                         6
Arithmetic mean                5.615385
Geometric mean                 5.261941
Harmonic mean                  4.823322
```

Sorting and Ordering

It is important to understand the distinction between sorting and ordering. Typically you have several variables in a dataframe, including the response variable and the various explanatory variables. In such a case, it is very dangerous to sort any one of the variables on its own, because it becomes uncoupled from its associated explanatory variables. This can cause terrible problems if statistical modelling is subsequently carried out, because values of the response variable will be associated with the wrong values of the explanatory variables. The answer is never to use sort on variables that are part of a dataframe.

```
sort(y)
```

```
[ 1]  2   3   4   5   5   5   6   6   7   7   7   8   8
```

produces the intuitively obvious output, as does

```
rev(sort(y))
```

```
[ 1]  8   8   7   7   7   6   6   5   5   5   4   3   2
```

The problems can arise if you say y < -sort(y) because there is no 'unsort' function. It is much better practice to leave the variables in your dataframe in their original unsorted sequence and to use order to produce new sequences, because this function leaves the original order undisturbed. Let's see it in action

```
order(y)
```

```
[ 1]  5   12  13   1   6   10   7   8   2   3   9   4   11
```

Now you will need to concentrate. What does the number 5 in the first element of order(y) mean? The thing you need to realize is that order (y) does **not** produce **values of** y. It produces **subscripts for** y. In particular, it produces the subscripts necessary to order the values of y into an increasing sequence. So the 5 in position 1 is a subscript – it says that the smallest value in y is the value in the fifth element of y. Let's see if that is correct:

```
y
```

```
[ 1]  5   7   7   8   2   5   6   6   7   5   8   3   4
```

Yes, it is. The smallest value in y is 2, and this is the fifth number in y. By the same logic, the eleventh value in y should be the largest, because 11 is the last number in order (y). This, too, is correct: there is a tie for highest number because there are two 8's, one in position 4 and one in position 11. Here are the values of z ordered by the matching elements of y

```
z[order(y)]
```

```
[ 1]  9   2   1   13   8   4   7   6   12   11  5   10   3
```

The fifth element of z is 9, then the twelfth element is 2, the thirteenth element is 1, the first is 13 and so on. The great advantage of order over sort is that it can be applied to whole dataframes, as illustrated on p. 20.

Counting Elements Within Arrays

You often want to know how many times particular values appear in a vector. The function for this is table. Here we generate 10 000 random numbers from a negative binomial distribution with mean $= 1.2$ and aggregation parameter $k = 0.63$ (prob $= k/($mu $+ k)$). The question is: how many zeros are there amongst the 10 000 numbers?

```
vals < -rnbinom(10000,size = 0.63,prob = 0.63/1.83)
table(vals)
```

```
vals
   0      1     2    3     4     5    6     7    8    9  10  11  12  13  14  15+
5048   2169  1116  651   379   230  156   97   59   31  29  13   9   5   3   5
```

The answer is 5048 zeros (about half of the numbers). You will get a different figure each time you execute the rnbinom function because the randomizations will be different.

Tables of Summary Statistics

One of the most commonly used functions in R is tapply. This is the function by which tables of means, variances, sample sizes and suchlike are produced. We need a big dataframe to see this in full swing:

```
Daphnia.data < -read.table("c:\\Daphnia.txt",header = T)
attach(Daphnia.data)
names(Daphnia.data)
```

```
[ 1] "Growth.rate"  "Water"  "Detergent"  "Daphnia"
```

There is a response variable (Growth.rate) and three categorical explanatory variables (Water, Detergent and Daphnia clone). The first argument of tapply is the variable to be summarized, the second argument is the variable by which the summary is to be classified, and the third argument is the function to be applied (mean, variance or whatever). Here is the use of tapply for means classified by the levels of a single categorical variable:

```
tapply(Growth.rate,Detergent,mean)
```

```
  BrandA        BrandB         BrandC          BrandD
3.884832      4.010044       3.954512        3.558231
```

```
tapply(Growth.rate,Water,mean)
```

```
   Tyne          Wear
3.685862      4.017948
```

When you want a two-way (or higher) classification, then the two (or more) classifying variables appear in a list as the second argument: the levels of the first variable create the rows (Water) and the levels of the second variable create the columns (Detergent) of the summary table:

tapply(Growth.rate,list(Water,Detergent),mean)

	BrandA	BrandB	BrandC	BrandD
Tyne	3.661807	3.911116	3.814321	3.356203
Wear	4.107857	4.108972	4.094704	3.760259

To check that the replication is equal for each combination of factors, you can use the length function as the third argument:

tapply(Growth.rate,list(Water,Detergent),length)

	BrandA	BrandB	BrandC	BrandD
Tyne	9	9	9	9
Wear	9	9	9	9

Yes, all combinations are based on nine numbers.

Converting Continuous Variables into Categorical Variables Using cut

It might be that you want to reduce a continuous variable into a categorical variable with a small number of levels (like small, medium and large). The cut function makes this very straightforward

```
sml < -cut(vals,3)
table(sml)
```

```
sml
(-0.017,5.66]    (5.66,11.3]    (11.3,17]
         9593            385           22
```

Here we take the vector of 10 000 negative binomial random numbers (p. 294) and use cut with table to see how many of them were small, medium or large (the elements of sml are produced by cutting vals into three equal parts based on the range of values in vals). There were only 22 individuals in the large category which ran from "[11.3" (meaning 11.3 and greater–'from and including'), to "17]" (meaning less than 17.0–'up to but not including'). Alternatively, instead of specifying the number of bits for cutting (three in this case), you can specify where you want the break points to be.

The split Function

The function called split produces a list of vectors on the basis of the levels of a factor, and is particularly useful in generating plots:

```
sdata < -read.table("c:\\temp\\splits.txt",header=T)
attach(sdata)
names(sdata)
```

```
[ 1] "xc" "yc" "fac"
```

The idea is to create scatterplots with different symbols for fac = "A" and fac = "B". Start by producing the blank

```
plot(xc,yc,type = "n",xlab = "x",ylab = "y")
```

Now create new vectors for the x and y axes split on the basis of fac = "A" or "B":

```
sxc < -split(xc,fac)
syc < -split(yc,fac)
```

add the points with different plotting symbols

```
points(sxc[[1]],syc[[1]])
points(sxc[[2]],syc[[2]],pch = 16)
```

then add the regression lines for each factor level separately:

```
for (i in 1:2) abline(lm(syc[[i]] ~ sxc[[i]]))
```

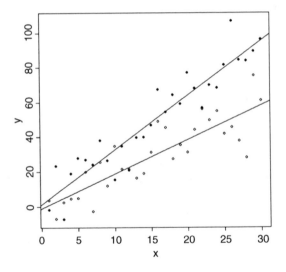

```
forms < -read.table("c:\\temp\\results.txt",header = T)
attach(forms)
names(forms)
```

```
[ 1] "return"   "bank"
```

The idea is to use split to produce a set of box and whisker plots, one for each bank

```
boxplot(split(return,bank),notch = T,ylab = "Return",xlab = "Bank")
```

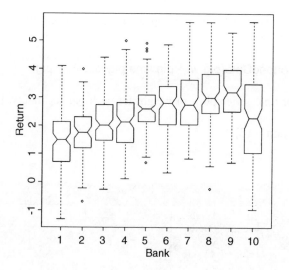

The notch = T option allows significance testing of the difference in median return between banks: those where the notches do not overlap are significantly different at 5% (like 4 and 5) while those where the notches do overlap (like 7 and 8) are not significantly different.

Trellis Plots

These multi-panel plots are particularly useful in the context of grouped data of the kind met in mixed effects modelling. Here are data on repeated measures on the growth of 48 pigs:

```
pigs < -read.table("c:\\temp\\pig.txt",header=T)
attach(pigs)
names(pigs)
```

```
[ 1] "Pig" "t1" "t2" "t3" "t4"  "t5" "t6"  "t7" "t8"  "t9"
```

```
pig.wt < -c(t1,t2,t3,t4,t5,t6,t7,t8,t9)
```

Next we create a vector for pig identity pig.id which is the vector Pig (a vector of numbers 1 to 48) repeated nine times

```
pig.id < -c(rep(Pig,9))
```

Now we generate a vector for the week number: 48 1s then 48 2s etc:

```
pig.time < -c(rep(c(1:9),each = 48))
pig < -data.frame(cbind(pig.time,pig.id,pig.wt))
```

Finally convert the dataframe into a grouped data object using groupedData

library(nlme)
pig.growth < -groupedData(pig.wt ~ pig.time|pig.id,data = pig)

To see a trellis of time series plots of weight for each pig separately, just type:

plot(pig.growth,pch = 16)

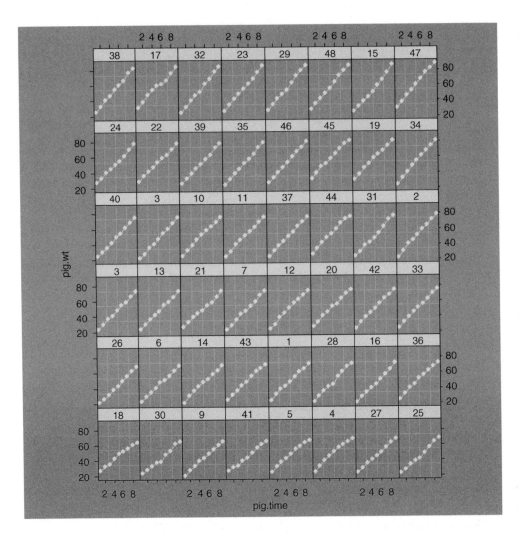

whereas to see all the time series in a single axis, set the outer option to ∼1 like this:

plot(pig.growth,outer = ∼ 1,key = F)

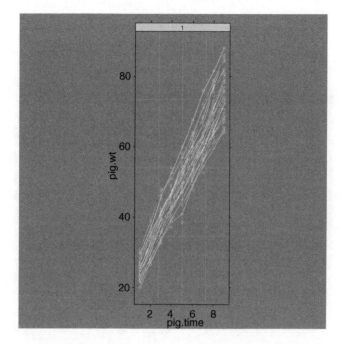

The lack of crossing of the growth trajectories shows that 'tracking' is high (the individuals that were the biggest at the beginning were the biggest at the end of the trial). Statistical modelling of these data is explained on the web site: http://www.imperial. ac.uk/bio/research/crawley/statistics.

The xyplot Function

Trellis plots are very useful for showing conditioning (the dependence of the response to one variable on the level of another continuous explanatory variable); they are accessible from the library called 'lattice':

```
library(lattice)
```

```
ozone < -read.table("c:\\temp\\ozone.data.txt",header = T)
attach(ozone)
names(ozone)
```

```
[ 1] "rad"  "temp"  "wind"  "ozone"
```

We want to look at the graph of ozone against wind-speed at different temperatures, using cut to produce six panels, based on the range of values of temp. The vertical bar, |, is read as 'given': we want to plot ozone against wind-speed, given the temperature.

```
xyplot(ozone$ozone ~ wind|cut(temp,6),
panel = function(x, y) {
```

```
panel.grid(h = -1, v =  2)
panel.xyplot(x, y,pch = 16)
panel.loess(x,y, span = 1) } )
```

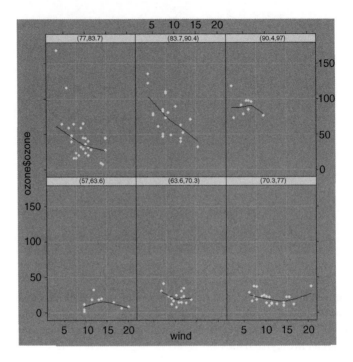

You can see that the dependence of ozone concentration on wind-speed is pronounced only for temperatures in the range 77–90°F. Note that when the dataframe has the same name as one of the variables contained within it, you need to extract the variable name using $ like this: ozone$ozone. Use the help function

?xyplot

to see all the options available in panel plots and trellis graphics.

Three-dimensional (3-D) Plots

For producing 3-D plots (including image and contour), the outer function is very useful. It evaluates a function (called func in this case) at every combination of x and y values within a square or rectangular array; this produces data in exactly the form required by image and contour. Note the use of add = T to add the contours on top of the coloured image.

```
x < -seq(0,10,0.1)
y < -seq(0,10,0.1)
```

```
func < -function(x,y) 3*x*exp(0.1*x)*sin(y*exp(-.5*x))
image(x,y,outer(x,y,func))
contour(x,y,outer(x,y,func),add = T)
```

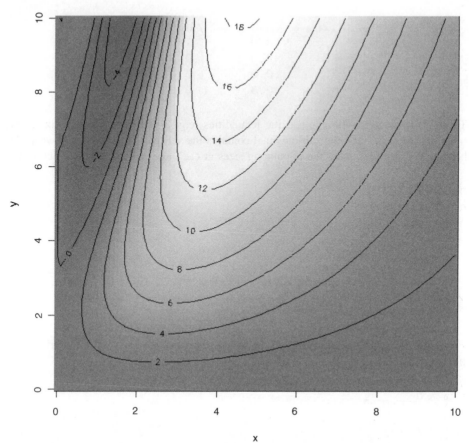

x

Other palettes for image plots include **terrain.colors** (greens, shading through yellows and browns into white for high ground) and **topo.colors** (blue, shading through greens to yellows for high ground) and **rainbow** (red, orange, yellow, green, blue, indigo, violet). For black and white printers you can specify a grey scale

```
image(x,y,outer(x,y,func),col = palette(gray(seq(0,.9,len = 25))))
```

Matrix Arithmetic

It is important to understand that matrix multiplication requires the %*% operator (not *). Consider this example where a Leslie matrix, **L**, is to be multiplied by a column matrix of age-structured population sizes, **n**

```
L < -c(0,0.7,0,0,6,0,0.5,0,3,0,0,0.3,1,0,0,0)
L < -matrix(L,nrow = 4)
```

Note that the elements of the matrix are entered in column-wise, not row-wise sequence. We make sure that the Leslie matrix is properly conformed:

```
L
```

	[,1]	[,2]	[,3]	[,4]
[1,]	0.0	6.0	3.0	1
[2,]	0.7	0.0	0.0	0
[3,]	0.0	0.5	0.0	0
[4,]	0.0	0.0	0.3	0

The top row contains the age-specific fecundities (e.g. 2-year-olds produce six female offspring per year), and the sub-diagonal contains the survivorships (70% of 1-year-olds become 2-year-olds). Now the population sizes at each age go in a column vector, **n**

```
n < -c(45,20,17,3)
n < -matrix(n,ncol = 1)
n
```

	[,1]
[1,]	45
[2,]	20
[3,]	17
[4,]	3

Population sizes next year in each of the four age classes are obtained by matrix multiplication, %*%

```
L %*% n
```

	[,1]
[1,]	174.0
[2,]	31.5
[3,]	10.0
[4,]	5.1

We can check this longhand. The number of juveniles next year (the first element of **n**) is the sum of all the babies born last year:

```
45*0 + 20*6 + 17*3 + 3*1
[1]  174
```

We write a function to carry out the matrix multiplication, giving next year's population vector as a function of this year's:

```
fun < -function(x) L%*%x
```

Now we can simulate the population dynamics over a period long enough (say, 40 generations) for the age structure to approach stability. So long as the population growth rate $\lambda > 1$ the population will increase exponentially, once the age structure has stabilized

```
pop < -numeric(40)
for (i in 1:40) {
  n < -fun(n)
  pop[i] < -sum(n)}
plot(log(pop),type = "l")
```

The population growth rate (the per-year multiplication rate, λ) is approximated by the ratio of population sizes in the 40th and 39th years:

```
pop[40]/pop[39]
```

```
[ 1]  2.164035
```

and the approximate stable age structure is obtained from the 40th value of **n**

```
n/sum(n)
```

```
              [ ,1]
[ 1,]    0.709769309
[ 2,]    0.230139847
[ 3,]    0.052750539
[ 4,]    0.007340305
```

The exact values of the population growth rate and the stable age distribution are obtained by matrix algebra: they are the dominant eigenvalue and eigenvector respectively. Use the function **eigen** applied to the Leslie matrix, **L**, like this

```
eigen(L)
```

```
$values
[ 1]  2.1694041+0.00000000i -1.9186627+0.000000000i -0.1253707+0.09751046i -0.1253707-
0.09751046i

$vectors
                  [ ,1]               [ ,2]                    [ ,3]                       [ ,4]
[ 1,]  -0.949264118+0i  -0.93561508+0i  -0.01336028-0.03054433i  -0.01336028+0.03054433i
[ 2,]  -0.306298338+0i   0.34134741+0i  -0.03616819+0.14241169i  -0.03616819-0.14241169i
[ 3,]  -0.070595039+0i  -0.08895451+0i   0.36511901-0.28398118i   0.36511901+0.28398118i
[ 4,]  -0.009762363+0i   0.01390883+0i  -0.87369452+0.00000000i  -0.87369452+0.00000000i
```

The dominant eigenvalue is 2.1694 (compared with our empirical approximation of 2.1640 after 40 years). The stable age distribution is given by the first eigenvector, which we need to turn into proportions

```
eigen(L)$vectors[,1]/sum(eigen(L)$vectors[,1])
```

```
[ 1]  0.710569659+0i  0.229278977+0i  0.052843768+0i  0.007307597+0i
```

This compares with our approximation (above) in which the proportion in the first age class was 0.70977 after 40 years (rather than 0.71057).

Solving Systems of Linear Equations

Suppose we have two equations containing two unknown variables:

$$3x + 4y = 12$$
$$x + 2y = 8.$$

We can use the function solve to find the values of the variables if we provide it with two matrices:

- a square matrix **A** containing the coefficients, and
- a column vector **kv** containing the known values.

We set the two matrices up like this (column-wise)

```
A <-matrix(c(3,1,4,2),nrow = 2)
A
```

	[,1]	[,2]
[1,]	3	4
[2,]	1	2

```
kv <-matrix(c(12,8),nrow = 2)
kv
```

	[,1]
[1,]	12
[2,]	8

Now we can solve the simultaneous equations

```
solve(A,kv)
```

	[,1]
[1,]	−4
[2,]	6

so $x = -4$ and $y = 6$ (as you can easily verify by hand). The function comes into its own when there are many simultaneous equations to be solved.

References and Further Reading

Agresti, A. (1990) *Categorical Data Analysis*. New York, John Wiley.

Aitkin, M., Anderson, D., Francis, B. and Hinde, J. (1989) *Statistical Modelling in GLIM*. Oxford, Clarendon Press.

Atkinson, A. C. (1985) *Plots, transformations, and Regression*. Oxford, Clarendon Press.

Bishop, Y. M. M., Fienberg, S. J. and Holland, P. W. (1980) *Discrete Multivariate Analysis: Theory and Practice*. New York, John Wiley.

Box, G. E. P. and Cox, D. R. (1964) An analysis of transformations. *Journal of the Royal Statistical Society, Series B* **26**: 211-246.

Box, G. E. P., Hunter, W. G. and Hunter, J. S. (1978) *Statistics for Experimenters: An Introduction to Design, Data Analysis and Model Building*. New York, John Wiley.

Box, G. E. P. and Jenkins, G. M. (1976) *Time Series Analysis: Forecasting and Control*. Oakland, California, Holden-Day.

Breiman, L., Friedman, L. H., Olshen, R. A. and Stone, C. J. (1984) *Classification and Regression Trees*. Belmont, California, Wadsworth International Group.

Caroll, R. J. and Ruppert, D. (1988) *Transformation and Weighting in Regression*. New York, Chapman and Hall.

Casella, G. and Berger, R. L. (1990) *Statistical Inference*. Pacific Grove, California, Wadsworth and Brooks/Cole.

Chambers, J. M., Cleveland, W. S., Kleiner, B. and Tukey, P. A. (1983) *Graphical Methods for Data Analysis*. Belmont, California, Wadsworth.

Chambers, J. M. and Hastie, T. J. (1992) *Statistical Models in S*. Pacific Grove, California, Wadsworth and Brooks Cole.

Chatfield, C. (1989) *The Analysis of Time Series: An Introduction*. London, Chapman and Hall.

Cochran, W. G. and Cox, G. M. (1957) *Experimental Designs*. New York, John Wiley.

Collett, D. (1991) *Modelling Binary Data*. London, Chapman and Hall.

Conover, W. J. (1980) *Practical Nonparametric Statistics*. New York, John Wiley.

Cook, R. D. and Weisberg, S. (1982) *Residuals and Influence in Regression*. New York, Chapman and Hall.

Cox, D. R. and Hinkley, D. V. (1974) *Theoretical Statistics*. London, Chapman & Hall.

Cox, D. R. and Oakes, D. (1984) *Analysis of Survival Data*. London, Chapman and Hall.

Cox, D. R. and Snell, E. J. (1989) *Analysis of Binary Data*. London, Chapman and Hall.

Crawley, M. J. (2002) *Statistical Computing: An Introduction to Data Analysis using S-PLUS*. Chichester, John Wiley.

Cressie, N. A. C. (1991) *Statistics for Spatial Data*. New York, John Wiley.

Crowder, M. J. and Hand, D. J. (1990) *Analysis of Repeated Measures*. London, Chapman and Hall.

Statistics: An Introduction using R   M. J. Crawley
© 2005 John Wiley & Sons, Ltd   ISBNs: 0-470-02298-1 (PBK); 0-470-02297-3 (PPC)

Dalgaard, P. (2002) *Introductory Statistics with R*. New York, Springer-Verlag.

Davidian, M. and Giltinan, D. M. (1995) *Nonlinear Models for Repeated Measurement Data*. London, Chapman and Hall.

Diggle, P. J. (1983) *Statistical Analysis of Spatial Point Patterns*. London, Academic Press.

Diggle, P. J., Liang, K.-Y. and Zeger, S. L. (1994) *Analysis of Longitudinal Data*. Oxford, Clarendon Press.

Dobson, A. J. (1990) *An Introduction to Generalized Linear Models*. London, Chapman and Hall.

Draper, N. R. and Smith, H. (1981) *Applied Regression Analysis*. New York, John Wiley.

Edwards, A. W. F. (1972) *Likelihood*. Cambridge, Cambridge University Press.

Efron, B. and Tibshirani, R. J. (1993) *An Introduction to the Bootstrap*. San Francisco, Chapman and Hall.

Everitt, B. S. (1994) *Handbook of Statistical Analyses Using S-PLUS*. New York, Chapman & Hall / CRC Statistics and Mathematics.

Ferguson, T. S. (1996) *A Course in Large Sample Theory*. London, Chapman and Hall.

Fisher, L. D. and Van Belle, G. (1993) *Biostatistics*. New York, John Wiley.

Fisher, R. A. (1954) *Design of Experiments*. Edinburgh, Oliver and Boyd.

Fleming, T. and Harrington, D. (1991) *Counting Processes and Survival Analysis*. New York, John Wiley.

Gordon, A. E. (1981) *Classification: Methods for the Exploratory Analysis of Multivariate Data*. New York, Chapman and Hall.

Gosset, W.S. (1908) writing under the pseudonym "Student". The probable error of a mean. *Biometrika* 6(1): 1–25.

Grimmett, G. R. and Stirzaker, D. R. (1992) *Probability and Random Processes*. Oxford, Clarendon Press.

Hairston, N. G. (1989) *Ecological Experiments: Purpose, Design and Execution*. Cambridge, Cambridge University Press.

Hampel, F. R., Ronchetti, E. M., Rousseeuw, P. J. and Stahel, W. A. (1986) *Robust Statistics: The Approach Based on Influence Functions*. New York, John Wiley.

Harman, H. H. (1976) *Modern Factor Analysis*. Chicago, University of Chicago Press.

Hastie, T. and Tibshirani, R. (1990) *Generalized Additive Models*. London, Chapman and Hall.

Hicks, C. R. (1973) *Fundamental Concepts in the Design of Experiments*. New York, Holt, Rinehart and Winston.

Hoaglin, D. C., Mosteller, F. and Tukey, J. W. (1983) *Understanding Robust and Exploratory Data Analysis*. New York, John Wiley.

Hochberg, Y. and Tamhane, A. C. (1987) *Multiple Comparison Procedures*. New York, John Wiley.

Hosmer, D.W. and Lemeshow, S. (2000) *Applied Logistic Regression*. 2nd Edition. New York, John Wiley.

Hsu, J. C. (1996) *Multiple Comparisons: Theory and Methods*. London, Chapman and Hall.

Huber, P. J. (1981) *Robust Statistics*. New York, John Wiley.

Huitema, B. E. (1980) *The Analysis of Covariance and Alternatives*. New York, John Wiley.

Hurlbert, S. H. (1984) Pseudoreplication and the design of ecological field experiments. *Ecological Monographs* 54: 187–211.

Johnson, N. L. and Kotz, S. (1970) *Continuous Univariate Distributions. Volume 2*. New York, John Wiley.

Kalbfleisch, J. and Prentice, R. L. (1980) *The Statistical Analysis of Failure Time Data*. New York, John Wiley.

Kaluzny, S. P., Vega, S. C., Cardoso, T. P. and Shelly, A. A. (1998) *S+ Spatial Stats*. New York, Springer-Verlag.

Kendall, M. G. and Stewart, A. (1979) *The Advanced Theory of Statistics*. Oxford, Oxford University Press.

Keppel, G. (1991) *Design and Analysis: A Researcher's Handbook*. Upper Saddle River, New Jersey, Prentice Hall.

Khuri, A. I., Mathew, T. and Sinha, B. K. (1998) *Statistical Tests for Mixed Linear Models*. John Wiley, New York.

Krause, A. and Olson, M. (2000) *The Basics of S and S-PLUS*. New York, Springer-Verlag.

Lee, P. M. (1997) *Bayesian Statistics: An Introduction*. London, Arnold.

Lehmann, E. L. (1986) *Testing Statistical Hypotheses*. New York, John Wiley.

Mandelbrot, B. B. (1977) *Fractals, Form, Chance and Dimension*. San Francisco, Freeman.

Mardia, K. V., Kent, J. T. and Bibby, J. M. (1979) *Multivariate Statistics*. London, Academic Press.

McCullagh, P. and Nelder, J. A. (1989) *Generalized Linear Models*. 2nd Edition. London, Chapman and Hall.

McCulloch, C. E. and Searle, S. R. (2001) *Generalized, Linear and Mixed Models*. New York, John Wiley.

Michelson, A. A. (1880) Experimental determination of the velocity of light made at the U.S. Naval Academy, Annapolis. *Astronomical Papers* 1: 109-145

Millard, S. P. and Krause, A. (2001) *Using S-PLUS in the Pharmaceutical Industry*. New York, Springer-Verlag.

Miller, R. G. (1981) *Survival Analysis*. New York, John Wiley.

Miller, R. G. (1997) *Beyond ANOVA: Basics of Applied Statistics*. London, Chapman and Hall.

Mosteller, F. and Tukey, J. W. (1977) *Data Analysis and Regression*. Reading, Mass., Addison-Wesley.

Nelder, J. A. and Wedderburn, R. W. M. (1972) Generalized Linear Models. *Journal of the Royal Statistical Society, Series A* **135**: 37–384.

Neter, J., Kutner, M., Nachsteim, C. and Wasserman, W. (1996) *Applied Linear Statistical Models*. New York, McGraw-Hill.

Neter, J., Wasserman, W. and Kutner, M. H. (1985) *Applied Linear Regression Models*. Homewood, Illinois, Irwin.

OED (2004) *Oxford English Dictionary*. Oxford, Oxford University Press.

O'Hagen, A. (1988) *Probability: Methods and Measurement*. London, Chapman and Hall.

Pinheiro, J. C. and Bates, D. M. (2000) *Mixed-effects Models in S and S-PLUS*. New York, Springer-Verlag.

Platt, J. R. (1964) Strong inference. *Science* **146**: 347–353.

Priestley, M. B. (1981) *Spectral Analysis and Time Series*. London, Academic Press.

Rao, P. S. R. S. (1997) *Variance Components Estimation: Mixed Models, Methodologies and Applications*. London, Chapman and Hall.

Riordan, J. (1978) *An Introduction to Combinatorial Analysis*. Princeton, New Jersey, Princeton University Press.

Ripley, B. D. (1996) *Pattern Recognition and Neural Networks*. Cambridge, Cambridge University Press.

Robert, C. P. and Casella, G. (1999) *Monte Carlo Statistical Methods*. New York, Springer-Verlag.

Rosner, B. (1990) *Fundamentals of Biostatistics*. Boston, PWS-Kent.

Ross, G. J. S. (1990) *Nonlinear Estimation*. New York, Springer-Verlag.

Santer, T. J. and Duffy, D. E. (1990) *The Statistical Analysis of Discrete Data*. New York, Springer-Verlag.

Searle, S. R., Casella, G. and McCulloch, C. E. (1992) *Variance Components*. New York, John Wiley.

Shao, J. and Tu, D. (1995) *The Jacknife and Bootstrap*. New York, Springer-Verlag.

Shumway, R. H. (1988) *Applied Statistical Time Series Analysis*. Englewood Cliffs, New Jersey, Prentice Hall.

Silvey, S. D. (1970) *Statistical Inference*. London, Chapman and Hall.

Snedecor, G. W. and Cochran, W. G. (1980) *Statistical Methods*. Ames, Iowa, Iowa State University Press.

Sokal, R. R. and Rohlf, F. J. (1995) *Biometry: The Principles and Practice of Statistics in Biological Research*. San Francisco, W.H. Freeman and Company.

Sprent, P. (1989) *Applied Nonparametric Statistical Methods*. London, Chapman and Hall.

Taylor, L. R. (1961) Aggregation, variance and the mean. *Nature* **189**: 732–735.

Upton, G. and Fingleton, B. (1985) *Spatial Data Analysis by Example*. Chichester, John Wiley.

Venables, W. N. and Ripley, B. D. (2002) *Modern Applied Statistics with S-PLUS*. 4th Edition. New York, Springer-Verlag.

Venables, W. N., Smith, D. M. and the R Development Core Team (1999) *An Introduction to R*. Bristol, Network Theory Limited.

Weisberg, S. (1985) *Applied Linear Regression*. New York, John Wiley.

Wetherill, G. B., Duncombe, P., Kenward, M., Kollerstrom, J., Paul, S. R. and Vowden, B. J. (1986) *Regression Analysis with Applications*. London, Chapman and Hall.

Winer, B. J., Brown, D. R. and Michels, K. M. (1991) *Statistical Principles in Experimental Design*. New York, McGraw-Hill.

Zar, J. H. (1999) *Biostatistical Analysis*. Englewood Cliffs, New Jersey, Prentice Hall.

Index

Entries in bold are R functions

1 parameter "1" as the intercept, 109
1:6 generate a sequence 1 to 6, 282
= = ("double equals") logical EQUALS, 100,
    157, 166, 244
!= logical NOT EQUAL, 83
    for **barplot**, 244
    influence testing, 161, 201
    with subsets, 144
- remove a term from a model, 107
/ division, 288
/ nesting of explanatory variables, 107, 176
"**\n**" new line in output, with **cat**, 292
+ add a term to a model, 107
$ component selection, 123, 300, 303
%% modulo, 27
%*% matrix multiplication, 302
%in% nesting of explanatory variables, 108
& logical AND, 20
| logical OR, 20
| conditioning ("given"), 107, 298
( ) arguments to functions, 19
(a,b] from and including a, up to but not
    including b, 273, 295
* main effects and interaction terms in a model,
    107
* multiplication , 288
: generate a sequence; e.g. 1:6, 20, 282
: interaction between two explanatory variables,
    107
[[ ]] subscripts for lists, 291
[ ] subscripts, 19, 289

[a,b) include b but not a,
\\ double backslash in file paths, 17
^ for powers and roots, 28, 281
{ } in defining functions, 24
    in for loops, 42
<- gets operator, 5
< less than, 20
> greater than, 20
1st Quartile with summary, 19, 52
3D plots introduction, 300
3rd Quartile with summary, 19, 52
a intercept in linear regression, 125
a priori contrasts, 209
abline function for adding straight lines to a
    plots, 127
    after Ancova, 193
    in Anova, 156
    with a linear model as its argument, 130
abline(0,2) draw a line with $a = 0$ and $b = 2$,
abline(h=3) draw a horizontal line at $y = 3$,
abline(lm(y~x)) draw a line with a and b
    estimated from the linear model **y~x**, 146,
    152
abline(v=10) draw a vertical line at $x = 10$,
absence of evidence, 3
acceptance null hypothesis, 4
additivity mis-specification, 124
age effects longitudinal data, 180
age-at-death data using **glm**, 113
aggregation and randomization, 10
aggregation count data, 241
AIC Akaike's Information Criterion, 208
air pollution correlations, 95

Statistics: An Introduction using R M. J. Crawley
© 2005 John Wiley & Sons, Ltd ISBNs: 0-470-02298-1 (PBK); 0-470-02297-3 (PPC)